T0275788

LONDON MATHEMATICAL SOCIETY LECTURE NOTE SERIES

Managing Editor: Professor J.W.S. Cassels, Department of Pure Mathematics and Mathematical
Statistics, University of Cambridge, 16 Mill Lane, Cambridge CB2 1SB, England

The books in the series listed below are available from booksellers, or, in case of difficulty,
from Cambridge University Press.

London Mathematical Society Lecture Note Series. 186

Complexity: Knots, Colourings and Counting

D.J.A. Welsh
University of Oxford

CAMBRIDGE
UNIVERSITY PRESS

Published by the Press Syndicate of the University of Cambridge
The Pitt Building, Trumpington Street, Cambridge CB2 1RP
40 West 20th Street, New York, NY 10011-4211, USA
10 Stamford Road, Oakleigh, Melbourne 3166, Australia

First published 1993

Library of Congress cataloguing in publication data available

British Library cataloguing in publication data available

ISBN   0 521 45740 8

Transferred to digital printing 2000

Contents

Preface

These lecture notes are based on a series of lectures which I gave at the Advanced Research Institute of Discrete Applied Mathematics (ARIDAM VI) in June 1991.

The lectures were addressed to an audience of discrete mathematicians and computer scientists. I have tried to make the material understandable to both groups; the result is that there are introductions to topics such as the complexity of enumeration, knots, the Whitney/Tutte polynomials and various models of statistical physics.

The main thrust throughout is towards algorithms, applications and the interrelationship among seemingly diverse problem areas. In many cases I have only given sketches of the main ideas rather than full proofs. However, I have tried to give detailed references. I have assumed some familiarity with the basic concepts of computational complexity and combinatorics, but I have aimed to define anything nonstandard when it is first encountered. My notation in both cases corresponds to standard usage, such as Garey and Johnson (1979) and Bollobás (1979).

Since the lectures I have rewritten the notes to incorporate some of the new developments but the basic material is the essence of what was presented. Much of the work was done when I held a John von Neumann Professorship at the University of Bonn. I am very grateful for the opportunity this offered, and to my friends at the Forschungsinstitut für Diskrete Mathematik, where the facilities and atmosphere make it such a stimulating place to visit.

I thank Larry J. Langley of the Department of Mathematics and Computer Science at Dartmouth College who took notes of my ARIDAM lectures and Brenda Willoughby who spent so much time producing them. I am also very grateful to James Annan and Kyoko Sekine for their criticisms of those parts of the text which I have used in graduate lectures at Bonn and Oxford.

It is a pleasure also to acknowledge the assistance of many friends who have done their best to answer my many queries when writing these notes. I owe a particular debt to Geoffrey Grimmett, Tony Guttmann, François Jaeger, Mark Jerrum, Peter Kronheimer, Raymond Lickorish, James Oxley, Paul Seymour, Morwen Thistlethwaite, Dirk Vertigan and Stuart Whittington.

Finally, I must thank the organisers and sponsors of the ARIDAM meeting; the United States Air Force Office of Scientific Research under AFOSR-89-0512A and AFOSR-90-0008A, and most of all, Peter Hammer and Nelly Segal who spent so much time and effort in organising the meeting and making it a success.

Dominic Welsh
December 1992

1 The Complexity of Enumeration

1.1 Basics of complexity

The basic notions of computational complexity are now familiar concepts in most branches of mathematics. One of the main purposes of the theory is to separate tractable problems from the apparently intractable. Deciding whether or not $P = NP$ is a fundamental problem in theoretical computer science. We will give a brief informal review of the main concepts.

We regard a computational problem as a function, mapping inputs to solutions, (graphs to the number of their 3-vertex colourings for example). A function is *polynomial time computable* if there exists an algorithm which computes the function in a length of time (number of steps) bounded by a polynomial in the size of the problem instance. The class of such functions we denote by FP. If A and B are two problems we say that A is *polynomial time Turing reducible* to B, written $A \propto B$, if it is possible with the aid of a subroutine for problem B to solve A in polynomial time, in other words the number of steps needed to solve A (apart from calls to the subroutine for B) is polynomially bounded.

The difference between the widely used P and the class FP is that, strictly speaking, both P and NP refer to *decision* problems.

A typical member of NP is the following classical problem known as *SATISFIABILITY*, and often abbreviated to *SAT*.

SATISFIABILITY

Instance: Boolean formula in conjunctive normal form, $C_1 \wedge \ldots \wedge C_m$, where the C_i are clauses in the literals $x_1, ..., x_n, \bar{x}_1, ..., \bar{x}_n$.

Question: Is there an assignment of truth values to the variables x_i which satisfies the formula?

It is clear that *verifying* that a given assignment of values satisfies the clauses can be done in time which is a polynomial function of the input size.

This is the essence of NP; it is the class of decision problems which have solutions which can be *checked* in polynomial time. A formal definition will be given later but perhaps the easiest way to think of it and one which is particularly pertinent to counting problems is as follows.

Visualise a *nondeterministic Turing machine (ndtm)* as an ordinary Turing machine with an additional input which can contain a *certificate* or *witness*. This certificate will be used in the verification process, which is an

1

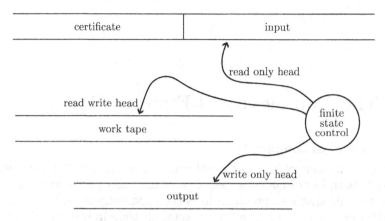

Figure 1.1. A nondeterministic Turing machine

ordinary Turing machine computation except that it is allowed to use the information in the certificate. This is the model used in Garey and Johnson (1979) and illustrated in Figure 1.1.

Thus in the case of *SATISFIABILITY* if the input x is a Boolean formula F, the certificate y might be a particular assignment of truth values and the working of the Turing machine is just the checking that with these values F is satisfied.

As another example, consider the language of composite integers. Formally this might be described by

COMPOSITE INTEGERS

Instance: Integer n expressed in binary.

Question: Is n composite?

The witness here might be a factorisation of n. The process of checking would be to verify that these factors multiplied together gave n, and this can be done in time which is bounded by a polynomial function of the size of the input, which in this case is $\lceil \log n \rceil$. Thus *COMPOSITE INTEGERS* belongs to NP. It is not obvious that the complementary language of *PRIME INTEGERS* belongs to NP. That it does, follows from a very nice paper of V. Pratt (1975) with the appealing title "Every prime has a succinct certificate".

The fundamental theorem of Cook (1971) is that NP has the hardest problems, known as the NP-complete languages. They are characterised by,

(i) $L \in NP$,

(ii) if there exists a polynomial time algorithm for deciding L, then $NP = P$.

In other words, L is NP-complete if it belongs to NP and for every other

$L' \in NP$, $L' \propto L$. There are now many hundreds of examples of *NP*-complete languages, including *SATISFIABILITY* and most of the difficult problems of graph theory such as deciding whether a graph G is k-colourable ($k \geq 3$) or has a Hamilton circuit.

A problem π is *NP-hard* if the existence of a polynomial time algorithm for π would mean there exists a polynomial time algorithm for some *NP*-complete language. Thus a language L is *NP*-complete if it is *NP*-hard and it belongs to *NP*.

Polynomial space, usually denoted by *PSPACE* consists of all languages recognisable by deterministic Turing machines which use an amount of space bounded by some polynomial function of the input size. It clearly contains *NP*, just run through all possible certificates, and the containment is thought to be strict.

Between *NP* and *PSPACE* is what is known as the *polynomial hierarchy*. This is the computational analogue of the Kleene arithmetic hierarchy of recursive function theory. A formal definition of it is found in Garey and Johnson (1979) but we can capture the set of languages in the polynomial hierarchy with the following definition. Define the class of languages *PH* by,

$$PH = \bigcup_{j=1}^{\infty} \Sigma_j^p$$

where the Σ_j^p are defined recursively by,

$$\Sigma_0^p = P,$$

$$\Sigma_{k+1}^p = NP^{\Sigma_k^p}$$

and where we are using standard notation for oracles. For a class Y of languages, NP^Y denotes the class of languages accepted by a nondeterministic polynomial time Turing machine with an oracle for any language in the class Y.

The containment relationships among these classes is shown in Figure 1.2, where *EXPTIME* denotes the class of decision problems solvable in time bounded above by $2^{p(n)}$ for some polynomial p.

Apart from knowing that P is a proper subset of *EXPTIME*, none of the other containments shown in Figure 1.2 is known to be strict.

1.2 Counting problems

The type of counting problem with which we shall be mainly dealing is of the following type.

For a given "universe", whether it be a d-dimensional lattice, graph, group or whatever, and a particular well defined "object" such as path, colouring, polygon, or automorphism on that universe, how many objects of given size are present?

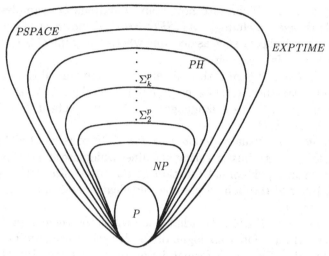

Figure 1.2.

In particular we shall be concerned with whether the associated counting problem can be done in time which is a polynomial function of the size of the universe or whether the problem is provably hard in some well defined sense.

We will not be concerned with problems such as the following.

(i) How many topologies are there on a set of size n?

(ii) How many nonisomorphic graphs are there on n vertices?

Nor will we be much concerned with the sort of enumeration problems covered in the monograph of Stanley (1986) where in most cases there is some hope of an answer in closed form.

Rather, we shall be concerned with the class $\#P$ of enumeration problems in which the objects being counted can be recognised in polynomial time. Since, if objects cannot be recognised it does not seem wise to try and count them we regard this property as justifying our calling $\#P$ the class of "sensible" counting functions.

Example. Let f be the function which maps any Boolean formula to its number of satisfying assignments. Then f is a member of $\#P$.

To see this, construct the ndtm M which, on input of a Boolean formula x, checks that a given assignment (or *witness* or *certificate*) y is an assignment of variables which makes x satisfiable. Thus M is a machine which can be interpreted as a function $g : \Sigma^* \times \Sigma^* \to \{0,1\}$, in such a way that

$$g(x,y) = \left\{ \begin{array}{ll} 1 & \text{if } y \text{ is a satisfying assignment to } x \\ 0 & \text{otherwise;} \end{array} \right\}$$

and also g is computable in polynomial time.   □

In other words $\#P$ is the class of *functions* that count the accepting paths of nondeterministic polynomial time Turing machines.

Valiant's original definition of $\#P$ in 1979 was as follows.

A *counting Turing machine* is a standard nondeterministic Turing machine with an auxiliary output device that (magically) prints, in binary notation, on a special tape the number of accepting computations induced by the input. It has (worst-case) *time complexity* $t(n)$ if the longest accepting computation induced by the set of all inputs of size n takes $t(n)$ steps (when the Turing machine is regarded as a standard non-deterministic machine with no auxiliary device).

(1.2.1) Definition: $\#P$ is the class of functions which can be computed by counting Turing machines of polynomial time complexity.

Another (equivalent) definition is given below. Its structure is appealing because it highlights the structural similarities between $\#P$ and the decision classes NP and RP (random polynomial time).

Consider first the following definition of nondeterministic polynomial time NP.

(1.2.2) Definition: A language $L \in NP$ if there exists a polynomial p and a polynomial time algorithm, which computes for each input x and each possible certificate (= guess) (= witness) y of length $p(|x|)$ a value $R(x,y) \in \{0,1\}$ such that

(i) if $x \notin L$ then $R(x,y) = 0$ for all y,

(ii) if $x \in L$ then $R(x,y) = 1$ for at least one possible certificate y.

(1.2.3) Definition: The class $RP = $ *random polynomial time* is defined by replacing "at least one" in (ii) by "at least half of the possible certificates".

Then we can define $\#P$ to be the class of *functions* $f : \Sigma^* \to \mathbf{N}$, such that there exists $L \in NP$ and associated p, R as defined above, such that

$$f(x) = |\{y : R(x,y) = 1\}| \quad x \in \Sigma^*.$$

Yet another formulation of $\#P$ is the following.

(1.2.4) Definition: Given the computation tree of a nondeterministic polynomial-time machine M, let $acc_M(x)$ be the number of accepting computations of M on input x. Then

$$\#P = \{f : \Sigma^* \to \mathbf{N} \text{ such that } f = acc_M \text{ for some polynomial time}$$
$$\text{non deterministic machine } M\}.$$

The language class $P^{\#P}$

We emphasise that $\#P$ is a class of functions, whereas the more familiar classes P and NP are classes of languages or sets. Thus, strictly speaking, $\#P$ is not directly comparable with the more familiar classes P and NP. Hence it is often more convenient to consider the associated class of languages $P^{\#P}$ defined by

$$P^{\#P} = \{L : L \text{ can be recognised in polynomial time by a Turing}$$
$$\text{machine equipped with a } \#P \text{ oracle}\}.$$

Until the recent results of Toda (1989), to be discussed below, not much more was known about the class $P^{\#P}$ than that:

(1.2.5) $P^{NP} \subseteq P^{\#P} \subseteq PSPACE.$

Proof: An oracle which counts the number of satisfying assignments of a Boolean formula is obviously at least as strong as an oracle which decides whether a Boolean formula is satisfiable. Thus $P^{NP} \subseteq P^{\#P}$. To show that $P^{\#P} \subseteq PSPACE$ we note that an exhaustive search and check of all possible satisfying assignments can be carried out in polynomial space. $\qquad\square$

1.3   # P-complete problems

We now define the class of $\#P$-hard (complete) functions in a way which is completely analogous to the familiar concepts NP-hard (complete).

Recall that FP is the class of functions which can be computed by deterministic polynomial time Turing machines.

(1.3.1) Definition: A problem π is *$\#P$-hard* if $\#P \subseteq FP^\pi$; it is *$\#P$-complete* if it is $\#P$-hard and it belongs to $\#P$.

Having defined $\#P$ and the concept of completeness and hardness for this class, it remains to exhibit some natural members. An obvious approach is to mimic the way in which NP-complete (hard) languages were found; this works, though as we shall see, a certain care has to be taken.

The strategy is to start with the counting counterpart of the generic NP-complete problem introduced by Cook (1971) in his seminal paper, and then to proceed in the same way as in NP-reductions but to ensure that the reductions in question preserve the number of solutions.

Accordingly we define:

 # TM-COMP: (Turing machine computation.)

Instance: Polynomial time nondeterministic Turing machine M, together with input $x \in \Sigma^*$.

Question: How many accepting computations has $M(x)$?

(1.3.2) Theorem: *# TM-COMP* is $\#P$-complete.

Proof. First show $\# \ TM\text{-}COMP \in \#P$. To do this we need to exhibit a nondeterministic machine N such that given any pair $\langle M, x \rangle$, with M a ndtm and x an input to M, will verify that M accepts x. This is clear, since N just simulates the computation $M(x, y)$ where y is a witness to the acceptability of x. Since M is polynomially bounded, the length of the computation is bounded by a polynomial in $|x|$.

Now consider any other function in $\#P$, call it g. Then by definition there exists a nondeterministic machine M_g such that for all $x \in \Sigma^*$,

$$|\{y : M_g(x, y) = 1\}| = g(x).$$

Thus clearly
$$\#P \subseteq FP^{\#TM\text{-}COMP}.$$

$\square$

We can now quickly build up a collection of $\#P$-complete problems as follows.

Take one of the standard NP-complete problems such as *SATISFIABIL-ITY (SAT)* or *HAMILTON CIRCUIT* and define $\# \ SAT$ and $\# \ HAMIL-TON \ CIRCUIT$ as follows:

SAT
Instance: Boolean formula F in conjunctive normal form.
Question: How many satisfying assignments has F?

HAMILTON CIRCUIT
Instance: Graph G.
Question: How many Hamilton circuits does G have?

(1.3.3) $\#$ SAT is $\#P$-complete.

Proof: It clearly belongs to $\#P$ and Cook's original reduction which showed that SAT is NP-complete is easily checked to preserve the number of solutions. $\square$

As pointed out first by Simon (1977) and Valiant (1979a), it seems to be the case that between most of the standard NP-complete languages there exist polynomial time transformations which preserve the number of solutions. In this way the original NP-completeness transformations developed by Karp (1972) show that the "natural" associated counting problem is also $\#P$-complete. Thus by tracing these transformations (and occasionally modifying them) we get for example:

(1.3.4) $\# \ HAMILTON \ CIRCUIT$ is $\#P$-complete.

(1.3.5) $\#k\text{-}COLOURINGS$ is $\#P$-complete for $k \geq 3$.

We illustrate the proof technique by showing the easy reduction between *SAT* and *3SAT*-(instances of *SATISFIABILITY* in which each clause has 3 literals).

(1.3.6) # *3-SAT* is #*P*-complete.

Proof: Suppose F has a clause with $i \geq 4$ literals. In this clause replace any two of these literals, say $x_j, \bar{x}_k$, by a new variable y and form

$$F' = F \wedge C[(x_j \vee \bar{x}_k) \equiv y]$$

where $C[(x_j \vee \bar{x}_k) \equiv y]$ is the conjunctive normal form for $(x_j \vee \bar{x}_k) \equiv y$.

Then F' has exactly the same number of solutions as F and the added clauses will have at most 3-literals per clause. The result follows by induction. □

Thus, the reduction above is a *parsimonious* reduction, in that it preserves the number of solutions unlike the easier Turing reduction between *SAT* and *3-SAT* given in Cook's original paper.

A slightly tedious, but probably worthwhile exercise is to go through the reductions between the classical *NP* complete problems and to check whether they preserve the number of solutions.

For those which are not, it is normally not too difficult to produce modified reductions which have this property. As a result we are able to make the following slightly vague claim.

(1.3.7) The counting versions of the classical *NP*-complete problems seem usually to be #*P*-complete.

However we urge caution here, there is no canonical definition of "classical" nor of the concept of a "natural" *NP*-complete problem. It is certainly not known to be true "that the counting counterpart of all natural *NP*-complete problems are #*P*-complete".

As a further note of caution we emphasise that the idea of "natural counting problem associated with an *NP*-complete language" is *not* a well defined concept; consider, for example, the decision problem *HAMILTON CIRCUIT*.

The "natural" nondeterministic Turing machine M for testing whether G is Hamiltonian will use as a certificate of membership a Hamilton circuit of G and check its existence.

However an equally good nondeterministic machine M' will take a path of length say $n-[\log n]$ in G (where $n = |V(G)|$) and check (by exhaustive search) whether it can be extended to a Hamilton circuit. This also can be done in polynomial time and (apart from at the intuitive level) it is not easy to say why M' is less natural than M.

Despite this caveat we shall (wherever we think there is no confusion) adopt the notation $\#(A, x)$ to denote the number of "natural" witnesses to

x belonging to the language A. Thus for example, $\#(SAT, F)$ will denote the number of assignments to the variables of F which make it satisfiable.

At this stage we formally define a notion which has been pervasive in this discussion.

(1.3.8) Definition: A *parsimonious transformation* from problem A to problem B is a polynomial time transformation f such that if $\#(A, x)$ denotes the number of solutions of problem A with instance x, then $\#(A, x) = \#(B, f(x))$.

Berman and Hartmanis (1977) have used padding arguments to show the existence of parsimonious transformations between most of the natural search problems associated with the classical NP-complete sets. As mentioned earlier Simon (1977) and Valiant (1979a) noted that the reductions between the early classical NP-complete sets were (or could be easily modified so as to be) parsimonious. However, as we shall show in a later section, it is *not* true that there are parsimonious reductions between all pairs of natural NP-complete languages.

It turns out that very few of the counting problems which arise naturally are known to have polynomial time algorithms. Indeed more can be said, in that most counting problems seem to be hard (or complete) for the class $\#P$. Thus they will not have polynomial time algorithms unless $\#P = FP$ and this would be much more surprising than $NP = P$.

We close this introduction by listing some of the, relatively few, nontrivial counting problems which are known to have P-time algorithms.

(1.3.9) Spanning trees of a graph. Counting the number of spanning trees in an undirected graph G reduces to evaluating any cofactor of the Kirchhoff matrix of G.

(1.3.10) Spanning arborescences of a graph. A *rooted arborescence* of a digraph Γ is a subgraph H which when undirected is a tree and which has a distinguished vertex v, the *root*, such that all edges of H are directed away from v. The number of spanning arborescences originating at any root of Γ is given by $T(\Gamma) = \det K_r(\Gamma)$ where r is any vertex and $K_r(\Gamma)$ is the Kirchhoff matrix of Γ.

(1.3.11) Eulerian cycles. The number of Eulerian cycles in a digraph Γ with vertex set $\{1, ..., m\}$ is given by

$$C(\Gamma) = \prod_{j=1}^{m} (d_j^+ - 1)! \, T(\Gamma)$$

where d_j^+ is the outdegree of vertex j and $T(\Gamma)$ is the number of spanning arborescences originating at a vertex.

(1.3.12) Perfect matchings of a planar graph. Kasteleyn (1967) gives a polynomial-time algorithm which computes the number of perfect matchings of any planar graph. This is in terms of a Pfaffian and is based on his earlier solution of the dimer problem in classical statistical physics.

The method also extends to the classes of graphs a) not containing $K_{3,3}$ as a minor and b) what are described as Pfaffian graphs. However, since it is not known how to decide if G is Pfaffian this extension has limited use. For a beautiful discussion of this problem see Lovász and Plummer (1986).

1.4   Decision easy, counting hard

It is not difficult to find examples of problems where finding a solution is very easy but counting the number of solutions is $\#P$-hard. Here are some examples of varying degrees of interest.

UNSAT

Instance: Boolean formula F in conjunctive normal form.
Question: How many assignments of the variables make F false?

(1.4.1) *# UNSAT* is $\#P$-complete.

Proof: It is clearly in $\#P$. But if there are n variables in F, the number of unsatisfying assignments when subtracted from 2^n gives an algorithm for *#SAT*.   □

A less trivial example is the following:

MONOTONE BOOLEAN FORMULA

Instance: Monotone Boolean formula F (that is it contains no negated variable).
Question: How many satisfying assignments has F?

Clearly finding a satisfying assignment to a monotone Boolean formula is trivial. However the counting problem is hard as we now see.

(1.4.2) *# MONOTONE BOOLEAN FORMULA* is $\#P$-complete.

Proof. Any Boolean formula F may be rewritten in the form

$$F = G \wedge H$$

where G and H are both monotone.

Then, counting the number of solutions of $G \wedge H$ and of G gives the number of solutions of F.

The rewriting of F consists of replacing each unnegated variable x_i in F by a new variable y_i and each negated variable $\bar{x}_k$ by z_k and conjoining F with

$$(\wedge_i(y_i \vee z_i)) \wedge (\neg \vee_i (y_i \wedge z_i)).$$

Since the function obviously belongs to $\#P$ it is $\#P$ complete. □

Note: The transformation above is parsimonious.

Another example in the same spirit of triviality as $\#UNSAT$ is:

NONCLIQUE
Instance: Graph G on n vertices.
Output: The number of sets of $\lceil n/2 \rceil$-vertices which are not cliques in G.

Finding such a "non clique" is trivial - just look for an edge. Counting them is obviously equivalent to counting cliques of size $\lceil n/2 \rceil$ and is therefore $\#P$-hard.

Another example which will be made more use of later is the following:

#DNF
Instance: Boolean formula F in disjunctive normal form.
Question: What is the number of satisfying assignments of F?

It is trivial to find a solution to such an F, however counting such solutions is hard as we now prove.

(1.4.3) Counting the number of satisfying assignments to a Boolean formula in disjunctive normal form is $\#P$-complete.

Proof. Given an instance F of $\#SAT$, it is not difficult, using de Morgan's laws, to transform it into a formula of polynomially equivalent length for $\neg F$, the complement of F. Then note that the number of satisfying assignments of F plus the number of satisfying assignments of $\neg F$ equals 2^k where k is the number of variables occurring in F. □

All the examples discussed in this section have had a certain aura of unreality or triviality about them. In the next section we consider a more significant example, which was already of interest in its own right in diverse areas of mathematics.

1.5 The Permanent

The *permanent* of an $n \times n$ matrix A is denoted by $per(A)$ and is defined by

$$per(A) = \sum_{\sigma \in S_n} a_{1\sigma(1)}a_{2\sigma(2)} \cdots a_{n\sigma(n)},$$

where S_n denotes the symmetric group of order n. In other words the permanent has the same expansion as the determinant but there is no problem over signs. Curiously, whereas the determinant is easy to evaluate the striking result of Valiant (1979b) is the following

(1.5.1) Theorem. Evaluating $per(A)$ is #P-hard, even for matrices in which every entry is 0 or 1.

Before proving this we note another interpretation.

Let G be a (not necessarily simple) bipartite graph, with bipartition $\langle U, W \rangle$ and with $|U| = |W| = n$. Let A be the *biadjacency* matrix of G, that is A_{ij} equals the number of edges joining u_i to v_j. Then

$$per(A) = \Psi(G)$$

where $\Psi(G)$ is the number of perfect matchings in G.

Recall that a *matching* of G is a set X of edges which have the property that no two members of X share a common vertex. It is a *perfect matching* if every vertex of G is the endpoint of some edge of X.

Thus in order for G to have a perfect matching it is necessary that $|V(G)|$ be even. Deciding whether G has a perfect matching and if it has, finding one, can be done in polynomial time. Matching theory has a huge range of applications from optimisation theory to statistical physics; a comprehensive and beautifully written treatise on the subject is that of Lovász-Plummer (1986). Here it is just worth mentioning in particular the connection with the *dimer problem* of mathematical physics. Let $\mathcal{L}_n$ denote an $n \times n$ section of the toroidal square lattice.

Then let $\psi(n)$ denote the number of ways of placing dimers on the edges so that each vertex is covered by exactly one dimer.

A classical result of statistical physics due to Fisher (1961a) and Kasteleyn (1961) says that

$$\lim_{n \to \infty} \psi(2n)^{\frac{1}{2}n} = \exp(2\mathcal{G}/\pi)$$

where $\mathcal{G} = \sum_{k=0}^{\infty} (-1)^k (2k+1)^{-2} = 0.915965594\ldots$ (Catalan's constant).

This is about 10% lower than 2, which is the number of ways (per lattice site) in which a free dimer can be placed on the lattice.

We now turn to the proof of Theorem 1.5.1.

Proof. Sketch: The main steps of the proof are as follows.

(i) There is a function $g \in FP$ which maps propositional formulae in conjunctive normal form to matrices with entries from $\{-1, 0, 1, 2, 3\}$, and such that for all F,

$$per\, g(F) = 4^{2\ell(F) - c(F)} \# F$$

where $\ell(F)$ is the number of occurrences of literals in F, $c(F)$ is the number of clauses in F and $\#F$ is the number of assignments satisfying F.

(ii) There is a transformation h, computable in time polynomial in m and the order of the matrix, that maps matrices with entries from the set $\{0, 1, ..., m\}$ to $(0, 1)$-matrices, such that for all A,

$$per(A) = per\, h(A).$$

(iii) If A is an $n \times n$ integer matrix with each entry bounded in magnitude by μ, then

$$|per(A)| \leq \mu^n n!$$

(iv) To compute $per(A)$ it is sufficient to compute its value *mod* p_i for each p_i in some set $\{p_1, ..., p_t\}$ of distinct prime numbers satisfying

$$p_1 p_2 \ldots p_t > 2\mu^n n!$$

This is a useful technique in counting problems. It is basically a special case of the Chinese remainder theorem.

(v) For any prime p, if A *mod* p is the matrix with (i, j) entry equal to A_{ij} *mod* p then $per(A)$ *mod* $p = per(A$ *mod* $p)$.

Combining (i)-(v) proves the theorem.   □

As an immediate corollary, if we regard computing as permanent as counting perfect matchings (see the dimer problem described earlier), we get:

(1.5.2) Counting perfect matchings in a bipartite graph is $\#P$-complete.

When it was first proved, this aroused interest because it was a first nontrivial example where the decision and construction problem was easy (that is in P) but the associated counting problem was hard. Since then there have been found many other examples of this phenomenon. Perhaps two of the simplest, in the sense that existence is absolutely trivial, are the following.

(1.5.3) Counting the number of forests in a bipartite planar graph is $\#P$-complete.

Note: a forest is any subset of edges containing no cycle, thus they always exist.

(1.5.4) Counting the number of linear extensions of a finite partial order is $\#P$-complete.

Note: again every partial order has at least one linear extension.

1.6   Hard enumeration problems not thought to be #P-complete

Not every hard counting problem is captured by the class #P. Consider for example:

HAMILTONIAN SUBGRAPHS

Instance: Graph G.

Output: Number of Hamiltonian subgraphs of G.

The status of this problem in the complexity hierarchy is interesting and not clear.

First note that it is a straightforward exercise to show:

(1.6.1) # *HAMILTONIAN SUBGRAPHS* is #P-hard.

However it does not appear to be #P-complete. The reason for this is that it does not seem easy to show that it is in #P. The "obvious" method fails as the natural "certificate" of the associated decision problem is a pair $\langle H', C \rangle$ where H' is a subgraph of G and C is a Hamilton circuit of H'. The *number* of such certificates is *not* the number of Hamiltonian subgraphs of G.

These are problems which are at least as hard as #P-complete problems. We now turn to a counting problem which is hard but which probably isn't NP- hard.

One of the outstanding open problems in the theory of NP-completeness is whether it is NP-hard to decide if two graphs are isomorphic.

The associated counting problem is:

ISOMORPHISM

Instance: Graphs G, H.

Output: Number of isomorphisms between G and H.

(1.6.2) Theorem (Mathon 1979) # *ISOMORPHISM* $\in FP^{NP}$.

More precisely, Mathon shows that equipped with an oracle to decide if two graphs are isomorphic, one can count isomorphisms in polynomial time.

This shows that counting isomorphisms is not significantly harder than deciding isomorphism. This could be regarded as further evidence that the isomorphism question is not NP-hard.

The class #NP

This is a class of functions introduced by Valiant in 1979 but which until recently seems to have received little attention. It is defined as follows.

Given a ndtm M and a set A define

$$acc_{M(A)} : \Sigma^* \to N$$

to be the function which, for arguments $x \in \Sigma^*$, takes a value equal to the number of accepting computations of M *with oracle A* when the input to M is x.

(1.6.3) Definition : $f \in \#NP \Leftrightarrow f = acc_{M(A)}$ for some language $A \in NP$ and M polynomially bounded.

The prototype member of $\#NP$ is the following.

#NSAT
Instance: Boolean formula F in variables $x_1, ..., x_n$ together with an assignation of values to x_i for $i \in J$ where $J \subseteq \{1, ..., n\}$.
Output: The number of assignments of values to the remaining variables which make F satisfiable.

Valiant pointed out that:

(1.6.4) $\#NSAT$ is complete for $\#NP$.

This is the typical problem of the class which we call *counting extendible configurations*. Another is the following.

N3COL
Instance: Graph G and a subset $X \subseteq V(G)$.
Question: How many 3-colourings of X are there which can be extended to a full 3-colouring of G?

Clearly this is $\#P$-hard; since an algorithm for doing this will give an algorithm for counting 3-colourings (take $X = V(G)$).

However it is not at all clear that this function belongs to $\#P$. Probably it does not, for what we can show is:

(1.6.5) Counting extendible 3-colourings is complete for $\#NP$.

Proof: The standard transformation between 3-colourability and *SAT* is parsimonious and this suffices since *NSAT* is complete for $\#NP$.   □

Now it is trivial that

(1.6.6) $\#P \subseteq \#NP$.

To see this just take M with no oracle in the definition of $\#NP$. It turns out that a consequence of some work of Toda (1989), which we discuss later, is that the gap between $\#P$ and $\#NP$ is not as great as was imagined. We return to this in §1.8.

1.7   Self-avoiding walks

Many questions in statistical physics are of the following type.

Let $\mathcal{F}$ be some well described family of patterns or configurations on a d-dimensional lattice $\mathcal{L}$. Let $\mathcal{F}_n$ be the set of members of $\mathcal{F}$ having size n – (here size is measured in some well recognised format). Then a typical problem is of the following type:

(i) How many patterns are there in $\mathcal{F}_n$?

This is usually much too hard for even moderate values of n. The lesser problem is:

(ii) Find the limiting value of some suitably normalised function of $\mathcal{F}_n$, such as $|\mathcal{F}_n|^{\frac{1}{n}}$, as $n \to \infty$.

Even this is usually too hard, though there have been some notable results such as the dimer solution mentioned in §1.5. Here we illustrate this idea with one well known classical problem of statistical physics and polymer chemistry, namely that of counting the number of self avoiding walks or polygons having n steps and rooted at the origin. For example, consider the square lattice with origin 0 and let c_n denote the number of self avoiding walks starting at 0 and having exactly n edges. Thus $c_1 = 4$, $c_2 = 12, \ldots$ and in general it is known that

(1.7.1) As $n \to \infty$, $c_n = e^{\kappa n + O(\sqrt{n})}$

where κ is known as the *connective constant* of the lattice. An exact value of κ is not known, though there are very accurate numerical estimates.

On any reasonable interpretation of "difficulty" this should be regarded as a hard problem, and of course in a sense it is. For we know that $f(n)$ is exponential in n and hence needs exponential time in order just to print the answer. In order to make the question more sensible we stipulate that the input should be an integer in *unary*.

Suppose we present this as a computational problem in standard format.

SELF-AVOIDING WALK

Instance: Integer n in unary.

Output: Number of self-avoiding walks on the square lattice having length n.

(1.7.2) *SELF-AVOIDING WALK* is contained in $\#P$.

Proof. Consider the ndtm M which accepts lattice paths if they are self-avoiding. This works in time polynomial in n in unary and therefore the associated counting function belongs to $\#P$.

The question which we cannot answer is the following:

(1.7.3) Problem: Is *SELF-AVOIDING WALK* complete in $\#P$?

I believe it is not, but obviously showing it is likely to be impossibly difficult.

The difficulty with discussing this sort of problem in the language of standard complexity is that "there are too few inputs". The temptation is to replace the lattice $\mathcal{L}$ by an arbitrary graph. But now we are just counting non-intersecting paths of length n in a graph and this reduces to $\# HAMILTON\ PATH$ and so is $\#P$-complete.

This led to the introduction by Valiant (1979a) of the peculiarly intractable class $\#P_1$. The typical problem of the class $\#P_1$ is of the following type. Let π be any property of graphs:

Input: Integer n in unary.

Question: How many graphs are there having property π and having n vertices?

Formally we make:

(1.7.4) Definition: $\#P_1$ is the class of functions which can be computed by counting Turing machines which have only a unary input alphabet and which have polynomial time complexity.

It is known that $\#P_1$ has complete members but it is fair to say that by any reasonable standard they are not natural. An explicit example of a naturally occurring complete problem for $\#P_1$ would be extremely interesting.

1.8   Toda's theorems

Until recently the relation between $\#P$ and its related counting classes and the more studied decision classes P, NP, PH was far from clear. For example, there was a folklore belief that a $\#P$-oracle was about as powerful as an oracle for a low level of the polynomial hierarchy such as Σ_3. This was shown to be almost certainly false by some remarkable results of Toda (1989).

Toda's proofs are difficult, they use a theory of operators which is still being developed, in particular see Toda and Watanabe (1992) and Köbler, Schöning and Toran (1989). Here we can do no more than outline the main ideas.

First we need to introduce the probabilistic class PP. This is the set of languages accepted in polynomial time by a Turing machine equipped with the ability to toss a fair coin. It is easy to see that

(1.8.1) $NP \subseteq PP \subseteq PSPACE.$

Furthermore, it is not difficult to show that a PP oracle is roughly equivalent in strength to a $\#P$-oracle. In other words,

(1.8.2) $FP^{\#P} = FP^{PP}$,

which means that any function computable in polynomial time when provided with an oracle for $\#P$ is also computable in polynomial time by a machine equipped with an oracle for PP. Toda's first theorem is the following.

(1.8.3) Theorem. The polynomial hierarchy is contained in P^{PP}.

An immediate corollary is that PP is at least as powerful as an oracle for any language in the polynomial hierarchy. This in turn, in view of (1.8.3) means that the following holds.

(1.8.4) Corollary. $FP^{PH} \subseteq FP^{\#P}$.

In other words:

(1.8.5) Any $\#P$-complete function is at least as difficult to compute as any problem from the polynomial hierarchy.

Thus the difficulty of computing a $\#P$-complete function is somewhere between $PSPACE$ complete languages and the hardest PH-functions in the hierarchy of difficulty. Thus, to describe a problem as $\#P$-complete must mean it is extremely intractable.

A further consequence of Toda's theorems is the following statement:

(1.8.6) $FP^{\#P} = FP^{\#NP}$.

The inclusion one way is obvious. However if one considers the following question it is clear that the reverse conclusion is surprising.

(1.8.7) Question: Given an oracle for counting Hamilton cycles, construct a polynomial time algorithm for counting Hamiltonian subgraphs.

A second of Toda's theorems relates the difficulty of parity computations with the polynomial hierarchy and we will consider this in §7.6.

1.9 Additional notes

Classical texts on complexity are Garey and Johnson (1979) and Balcázar, Diaz and Gabarró (1988). An invaluable compact source is the catalogue of Johnson (1990). I first met the definitions of NP and RP as given in §1.2 in Lovász (1979b). Proofs of the counting results (1.3.9)–(1.3.11) can be found in Tutte (1984) and Lovász (1979a). Kasteleyn's (1967) algorithm for counting perfect matchings in the planar case is based on his (1961) paper counting dimer configurations (see also Fisher (1961a)). A very interesting account of the relation between the Pfaffian problem and other algorithmic questions is given in Lovász and Plummer (1986) and Thomassen (1992). The results (1.4.2) and (1.4.3) on the hardness of counting solutions to DNF and monotone Boolean formulae are from Valiant (1979a) which also contains several other examples of decision easy/counting hard problems.

The #P-completeness results (1.5.3) and (1.5.4) are from Vertigan and Welsh (1992) and Brightwell and Winkler (1991) respectively. The existence of the connective constant of self-avoiding walk theory is due to Hammersley (1957). The bound (1.7.1) is from Hammersley and Welsh (1962) but it is widely believed that the asymptotic formula $c_n = e^{\kappa n + O(\log n)}$ ought to hold, see Hammersley (1991). For more on the theory of self-avoiding walks see the remarkable paper of Hara and Slade (1991).

A somewhat simplified proof of Toda's theorem is given by Babai and Fortnow (1990) who also characterise #P-functions in terms of straight line programmes of multivariate polynomials.

For an account of recent developments in the structural complexity of counting classes see Schöning (1990).

2 Knots and Links

2.1 Introduction

Some of the first work on knot theory was motivated by its applications. For example an early result of Gauss (1833) concerned the magnetic field which a knotted wire would produce if it carried an electric current. Kelvin and others tried to understand atoms as knots hoping that the table of knots would reproduce the regularities of the table of elements.

In the last ten years there has been an upsurge of interest due in some part to other applications of knots in such subjects as biology (enzyme recognition), chemistry (polymer theory) and physics (statistical mechanics). Molecular knots have long been the subject of speculation and in 1989, Dietrich-Buchecker and Sauvage succeeded in synthesising the trefoil knot with molecular threads and thus produced the world's smallest knot.

Here I shall attempt to give an introduction to combinatorial knot theory with the emphasis on classification, applications and the complexity of recognition problems.

Classical knot theory is concerned with embeddings of a circle (1-sphere) S^1 into Euclidean 3-space $\mathbf{R}^3$ or the 3-sphere S^3, as illustrated.

If X, and Y are Hausdorff spaces a mapping $f : X \to Y$ is called an *embedding* if $f : X \to f(X)$ is a homeomorphism.

A *knot* is an embedding f of S^1 into S^3, though it is usually identified with its image $f(S^1)$. In other words a knot is a subset of $\mathbf{R}^3$ which is homeomorphic to a circle.

A *link* with k components is a subset of $\mathbf{R}^3$ which is homeomorphic to the disjoint union of k circles.

Trefoil   Figure-eight knot

Figure 2.1.

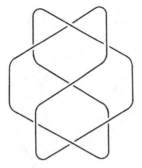 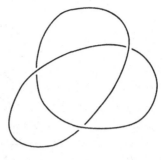

Figure 2.2.

Two knots K, L are *ambient isotopic* if there exists a homotopy h_t : $S^3 \to S^3$ ($0 \le t \le 1$) such that $h_0 = 1$, each h_t is a homeomorphism and $h_1(K) = L$.

As R.H. Fox in his paper entitled "A remarkable simple closed curve" (Ann. Math. 1949) pointed out, topological embeddings of $S^1 \to S^3$ may have strange features. In particular there exist knots which have no piece-wise linear representation. A knot K is said to be *tame* if it is ambient isotopic to a simple closed polygon in S^3. A knot is *wild* if it is not tame.

Henceforth when we refer to a knot we shall assume it is tame. Consequently we can always regard it as having a piecewise linear representation. With this representation in mind we can think of *ambient isotopy* as being any topological operation which deforms a polygon made up of vertices joined by "rubber band" edges and anything is allowed except cutting edges, subject to the proviso that distinct edges do not share points, except possibly end points, and vertices remain distinct.

A knot is usually described by a projection onto a plane. A projection of K is a *regular projection* if

(a) it contains only finitely many multiple points

(b) all multiple points are double points and these are transverse points (that is crossing points).

Given such a regular projection, provided each crossing point has specified which is the over/under crossing the projection determines the knot. Usually this information is conveyed by leaving a gap in the string of the undercrossing, this then constitutes the *knot diagram*.

Example. Two knot diagrams representing the same knot, the trefoil, are shown in Figure 2.2.

Two knot diagrams are called *equivalent* if they are connected by a finite sequence of moves shown in Figure 2.3 or their inverses. In each case, away from the local area of the move, the diagrams are unchanged. Any such move is called a *Reidemeister move*.

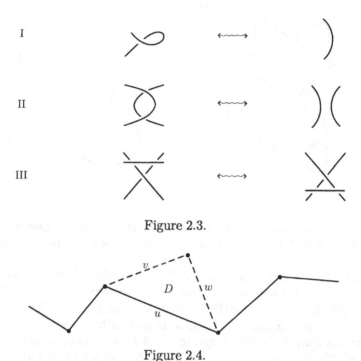

Figure 2.3.

Figure 2.4.

The fundamental theorem of Reidemeister (1935) is:

(2.1.1) Theorem. Two knots are ambient isotopic if and only if any diagram of one is equivalent to any diagram of the other under Reidemeister moves.

The proof of this fundamental theorem is based on the idea of knots being *combinatorially equivalent*. This concept, as described in Burde and Zieschang (1985), is in turn based on the notion of two polygonal knots K, L, in $\mathbf{R}^3$ being *related by a Δ-move* if L can be obtained from K by replacing a straight line segment u of K by the two other sides of a triangle D having sides u, v, w; it being assumed that K and D intersect only in this segment u. We illustrate this move in Figure 2.4.

If L is obtained from K by a Δ-move then the reverse process is a Δ^{-1}-*move*. Then say that two knots are *combinatorially equivalent* if one can be obtained from the other by a finite sequence of Δ- and Δ^{-1}-moves.

The major step in the proof of Reidemeister's theorem is proving:

(a) Two knots K, L are ambient isotopic if and only if they are combinatorially equivalent.

Then, it is relatively easy to show that:

+ve

-ve

Figure 2.5.

(b) Two regular projections of the same simple closed polygon are connected by a finite sequence of Reidemeister moves.

It remains only to prove:

(c) Each Δ- and Δ^{-1}-move induces Reidemeister moves on the projection.

For a full proof we refer to Burde and Zieschang (1985).

A link is *oriented* if each of its components is given an orientation. We shall be principally concerned with unoriented knots but point out that there is an oriented version of Reidemeister's theorem. In this case the allowable moves must include all possible orientations of the strings.

The two fundamental algorithmic problems about knots are a) the *unknotting problem*, namely how to decide whether a knot really is knotted and b) the *recognition problem*, namely, given two links how to decide whether they are (ambient) isotopic. Algorithms exist for both of these problems, see Hemion (1992), however their status in the complexity hierarchy seems very unclear.

2.2 Tait colourings

Given any link diagram D it is easy to prove that we can colour the faces black and white in such a way that no two faces with a common edge are the same colour. By convention, we colour the boundary faces black, and this is the *checkerboard* or *Tait colouring* of the diagram D.

From this we can get a canonical signed graph $S(D)$, its vertices are the black faces of the Tait colouring and two vertices are joined by a signed edge if they share a crossing. The sign of the edge is positive or negative according to the (conventional) rule shown in Figure 2.5.

Example. The Tait colouring and associated signed graph of the diagram D are illustrated in Figure 2.6.

Clearly $S(D)$ is a plane graph. Moreover, it is easy to prove that every signed plane graph can arise in this way. More precisely we have:

(2.2.1) Proposition : There is a 1-1 correspondence between link diagrams and signed plane graphs.

Proof. It remains only to show that given any signed plane graph S we can associate with it a link diagram D such that $S(D) = S$. Given any plane

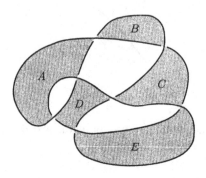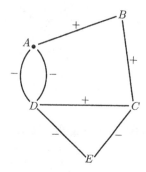

Figure 2.6.

graph G, the *medial graph* $m(G)$ has vertices at the midpoint of each edge of G and two vertices are joined by an edge in $m(G)$ if they are consecutive edges of a face (including the infinite face). Thus $m(G)$ is 4-regular. The vertices of $m(G)$ are going to be the crossings of the link diagram D; the sign of the underlying edge of S determines which string goes over/under in D.   □

In the same way as infinitely many link diagrams can represent equivalent links, for any link L there are infinitely many signed plane graphs which represent it.

It is routine to check that the Reidemeister moves (I) - (III) can be translated into the local moves on signed graphs shown in Figure 2.7.

Thus an equivalent formulation of Reidemeister's theorem is:

(2.2.2) Theorem. Two plane signed graphs S_1 and S_2 represent the same link if and only if S_1 can be transformed into S_2 by some finite sequence of the moves (I′) to (III′) and their inverses.

A very easy observation is the following.

(2.2.3) If G is any plane signed graph and G^* is its plane dual with the signs of the edges multiplied by (-1) then the links $L(G)$ and $L(G^*)$ are isotopic.

The proof is an easy consequence of the star triangle transformation; alternatively it can be seen by considering the two representations of the associated links on a sphere.

This leads naturally to the concept of the *mirror image* of a link L, defined in the obvious way to be that link $\bar{L}$ obtained from any link diagram D of L by interchanging the over/under nature of the strings at each crossing.

In other words, the signs of $S(D)$ are multiplied by (- 1). A link is called *amphicheiral* or *achiral* if it is isotopic to its mirror image.

No good (that is fast and infallible) test of amphicheirality is known.

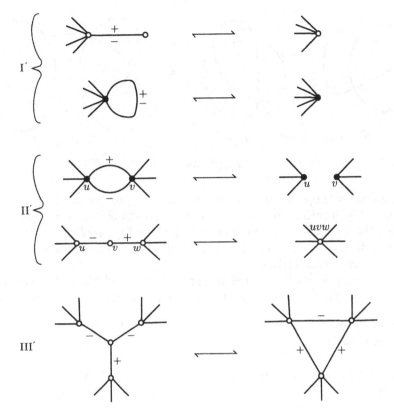

Figure 2.7.

2.3 Classifying knots

Two link diagrams which have different numbers of components obviously represent nonisotopic links. This is probably the simplest example of link invariant and is used as a first subdivision of links in the various tables of links, see for example Thistlethwaite (1985), Rolfsen (1976) or Burde and Zieschang (1985).

A less trivial invariant is the *crossing number* which is defined as the minimum number of crossings in any diagram representing the link. Determining the crossing number seems to be difficult; certainly no good (polynomial time) algorithm is known.

All knots with crossing number less than or equal to 7 have the property that they can be represented by a link-diagram in which the crossings are alternately over/under/over ... . Any link with this property is called *alternating*, they are an important, but relatively small subclass of the class of links. It is easy to see that:

(2.3.1) A link L is alternating iff it has a signed graph representation in which all edges have the same sign.

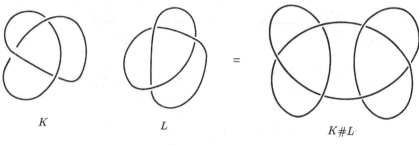

$$K \qquad\qquad L \qquad\qquad\qquad K\#L$$

Figure 2.8.

A fundamental operation on knots K_1 and K_2 is to "tie them together" and then to join their loose ends to form a closed string. The resulting knot is called the *product* of K_1 and K_2 and is denoted by $K_1\#K_2$.

A knot is *prime* if it is not the unknot and cannot be expressed as the product of two nontrivial knots.

In classifying knots it is usual to list only those which are prime, but again this has the drawback that it is not easy to decide if a knot is prime.

The knot complement

It is easy to see that :

(2.3.2) If two links are ambient isotopic then their complements are homeomorphic.

Proof. If K is ambient isotopic to K' under h then the restriction of h to the complement is a homeomorphism.   □

Also we have:

(2.3.3) A link and its mirror image have homeomorphic complements.

It is also possible for pairs of links which are some way from being ambient isotopic to have homeomorphic complements. This follows from a famous example of Whitehead (1937), who observed:

(2.3.4) If I_n ($n \in Z$) denotes the link with diagram shown in Figure 2.9 then I_{2n} has a complement homeomorphic to that of I_0 for all $n \in \mathbf{Z}$ but I_{2n} is *not* ambient isotopic to I_{2m} if $n \neq m$.

However Gordon and Luecke (1989) have obtained the very important result that for knots as against links the following is true.

(2.3.5) Theorem. If two knots K, K' in S^3 are such that their complements are homeomorphic then there exists a homeomorphism $h : S^3 \to S^3$ such that $h(K) = K'$.

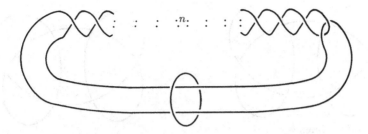

Figure 2.9.

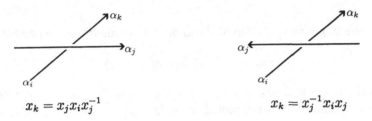

Figure 2.10.

This can be restated in the following form.

(2.3.6) If two knots have homeomorphic complements then they are equivalent under Reidemeister moves and taking mirror images.

The knot group

The *knot group* is the fundamental group of the knot complement. Its objects are the homotopy classes of paths in the knot complement, with the product of two closed paths α and β from an arbitrary point O in the complement, being the path γ consisting of α followed by β.

There is an easy way to construct and describe the knot group; we will first give the recipe to produce what is known as the *Wirtinger presentation* and then give the motivation.

An *arc* or *segment* of a knot is the portion of string which begins at one under crossing and continues until the next undercrossing. First note that if we start at some arbitrary crossing and label the segments as we traverse an oriented knot K, calling the successive segments α_i, then:

(2.3.7) A knot diagram with n crossings has exactly n segments.

We associate an element x_i of the knot group with each segment α_i, $1 \leq i \leq n$.

Moreover at each crossing there will be 3 segments as shown.

These induce the relation shown. Note that this relation depends only on the orientation of the jth segment and is based on the 'left hand screw' rule.

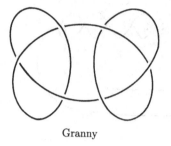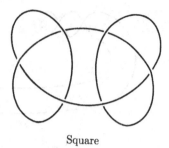

Granny

Square

Figure 2.11.

The Wirtinger presentation of the knot group consists of the presentation

$$\langle x_1, x_2, \ldots, x_n \mid r_1, r_2, \ldots, r_n \rangle$$

where the r_i is the relation at the ith crossing and is typically represented in the shorthand form exemplified by $x_k^{-1} x_j x_i x_j^{-1}$ $(= e)$ or $x_k^{-1} x_j^{-1} x_i x_j$ in the two cases above.

One of the relations in the Wirtinger presentation will always be redundant and one can play with different presentations, making substitutions and the like to find some non-obvious presentations. For example we leave it as an exercise to show:

(2.3.8) The trefoil has a presentation $\langle u, v; u^3 = v^2 \rangle$.

Again, like the knot complement, the knot group is a strong invariant. Obviously knots having homeomorphic complements will have the same group. However, the group is not as strong as the complement in the following sense.

(2.3.9) There exist knots with the same group but with different complements.

Example. Take the granny knot and the square knot shown in Figure 2.11. One is the product of the left and right trefoils, the other is the product of right trefoils. Their groups are therefore isomorphic but their complements can be shown to be not homeomorphic. □

However these knots are not prime; for prime knots we have what is known as Whitten's Rigidity Theorem.

(2.3.10) Theorem. Prime knots with isomorphic groups have homeomorphic complements.

This still leaves us with two big problems - how to test whether a knot is prime and a second problem which is of a group theoretical character. This is, given two finitely presented knot groups, how do we test whether they are isomorphic.

As far as the unknot question is concerned we can use the following fundamental theorem of Dehn-Papakyriakopoulos.

(2.3.11) Theorem. The trivial knot is the only knot having a single generator and no relations as its group.

However, even this does not settle the unknotting problem since there seems no easy way of determining whether a knot group is equivalent to that of the trivial knot.

Markov properties of groups

Let π be a property of finitely presented groups which is preserved under isomorphism. Then π is a *Markov property* if:

(i) there exists a finitely presented group with π;

(ii) there exists a finitely presented group which cannot be embedded in any finitely presented group which has π.

The following result of Adyan (1958) and Rabin (1958) shows that almost nothing about finitely generated groups is decidable.

(2.3.12) Theorem. If π is a Markov property then π is undecidable.

Proof Idea: Use the knowledge that the word problem for finitely generated groups is undecidable. Show that any Markov property algorithm would solve the word problem. □

However a deep theorem proved by Waldhausen in 1968 is the following.

(2.3.13) Theorem. The (fundamental) group G of any (tame) knot has a solvable word problem.

For a discussion of this and related problems see Waldhausen (1978).

2.4 Braids and the braid group

A *braid* on m strings is constructed as follows.

Take m distinct points $P_1, \ldots, P_m$ in a horizontal line and link them to distinct points $Q_1, \ldots, Q_m$ lying in a parallel line by m disjoint simple arcs (strings) f_i in $\mathbf{R}^3$, with f_i starting at P_i and ending at $Q_{\pi(i)}$ and where π is a permutation of $(1, 2, \ldots, m)$. The f_i are required to "run downwards" as illustrated in the example shown in Figure 2.12.

The collection of strings constitutes an *m-braid*. The map $i \mapsto \pi(i)$ is the *permutation* of the braid. The braid will be *closed* by joining the points P_iQ_i as illustrated in Figure 2.12(b).

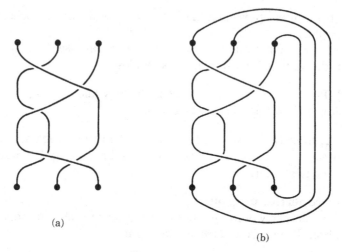

Figure 2.12.

(2.4.1) Each closed braid defines a link of μ components, where μ is the number of cycles in the permutation π.

The oriented link formed by closing the braid α will be denoted by $\hat{\alpha}$. The trivial m-braid is a configuration in which no crossing of strings occurs. For example Figure 2.12 shows a braid on 3 strings representing a link of two components and having crossing number ≤ 5. Hence trivially, every closed braid is a link. The converse also holds.

(2.4.2) Theorem (Alexander). *Every link can be represented as a closed braid.*

There is an obvious way in which braids on the same number of strings can be composed. Namely, if z is a braid having end points $Q_1, \ldots, Q_m$ and z' is a braid having initial points $P'_1, \ldots, P'_m$, their composition zz' is obtained by identifying Q_i with P'_i for $1 \leq i \leq m$; the resulting braid has initial points $P_1, \ldots, P_m$ and endpoints $Q'_1, \ldots, Q'_m$. It is straightforward to check:

(2.4.3) Under the above composition the isotopy classes of m-braids form a group, called the braid group B_m.

It is also clear that the braid group B_m is generated by the *elementary braids* σ_i, σ_i^{-1} $(1 \leq i \leq m-1)$ representing simple interchanges.
Defining relations for B_m were proved by Artin to be

$$\sigma_i \sigma_j = \sigma_j \sigma_i \qquad \text{if} \quad |i - j| \geq 2$$

and

$$\sigma_i \sigma_{i+1} \sigma_i = \sigma_{i+1} \sigma_i \sigma_{i+1}.$$

Figure 2.13.

Also, there exists a "Reidemeister type" theorem for braids due to A.A. Markov. This is in terms of moves of the following kind:

Markov moves

TYPE I: Replace braid $\alpha \in B_m$ by a conjugate $\gamma \alpha \gamma^{-1} \in B_m$ with $\gamma \in B_m$.

TYPE II: Replace $\alpha \in B_m$ by $\alpha \sigma_m \in B_{m+1}$ or $\alpha \sigma_m^{-1} \in B_{m+1}$.

TYPE II^{-1}: Replace a braid of the form $\alpha \sigma_m \in B_{m+1}$, respectively $\alpha \sigma_m^{-1} \in B_{m+1}$, by $\alpha \in B_m$, provided α is a word in the generators $\sigma_1, \ldots, \sigma_{m-1}$ only.

(2.4.4) Theorem. Two braids have closures which are equivalent as links if and only if they are connected by a finite sequence of elementary moves of type I, II and II^{-1}.

A proof of Markov's theorem is given in the book of Birman (1974).

A given link can be represented in infinitely many different ways over many different braid groups and if $\alpha \in B_m$ and $\beta \in B_n$ have isotopic closures then the sequence of Markov moves transforming α to β may be long and go through several different braid groups.

Pictorially, Markov's moves are easy to understand. The conjugacy relation represented by a type I move is nothing more than the observation that the closure of the braid $\gamma \alpha \gamma^{-1}$ is isotopic to $\hat{\alpha}$ since closing the braid allows γ^{-1} to cancel out the effect of γ.

The type II moves are the moves representing the introduction of a new string (or its inverse). However it is these moves which cause difficulties, because they change the number of strings and thus stop the problem being just a conjugacy problem. Although deciding conjugacy in a given braid group is difficult, there do exist algorithms due to Makanin (1968) and Garside (1969).

2.5   The braid index and the Seifert graph of a link

As we have seen, each link in 3-space has many different representations as a closed braid. The minimum number of strings in any braid representation of L is known as the *braid index* of L and is denoted by $\beta(L)$. In other words $\beta(L)$ is the smallest m for which there exists $\alpha \in B_m$ with $\hat{\alpha}$ isotopic to L.

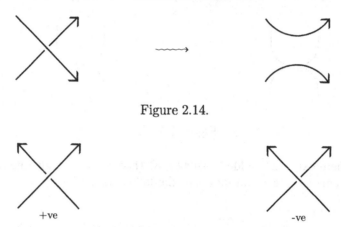

Figure 2.14.

Figure 2.15.

The braid index characterises the unknot in the following sense:

(2.5.1) K is the unknot iff $\beta(K) = 1$.

Thus any polynomial time algorithm which determines the braid index would be of great interest.

A classical theorem about links is the following.

(2.5.2) Any oriented link L is the boundary of a compact connected orientable surface.

A canonical way of constructing such a surface, was given by H. Seifert in (1934) and the resulting surface is known as a *Seifert surface*. The key step in the construction of a Seifert surface from an oriented link diagram D is to "split" each crossing of D in the obvious way shown in Figure 2.14 and then to glue the resulting set of disjoint discs together using twisted bands to preserve orientability. For more details see Kauffman (1987b).

The *genus* $g(K)$ of a knot is the minimum genus of an oriented surface Σ in S^3 which has K as its boundary. Thus, by taking any Seifert surface of a knot K and then determining its genus we can obtain an upper bound on $g(K)$. Moreover this can be done quickly (and easily). However determining $g(K)$ seems to be hard. It is certainly as hard as the unknotting problem since it is clear that $g(K) = 0$ if and only if K is equivalent to the unknot.

The *Seifert graph* $\Gamma(D)$ of an oriented link diagram D is a signed graph whose vertices are the Seifert circles (or discs) constructed in the above splitting process and with signed edges joining two circles whenever they share a crossing. The sign of the crossing is determined by the convention shown in Fig. 2.15.

An easy property of $\Gamma(D)$ is that:

(2.5.3) Any Seifert graph is planar and bipartite.

Proof (sketch). Planarity is obvious; to show that it is bipartite assume there is an odd circuit. This forces a contradiction on the clockwise/anticlockwise orientations of the Seifert circles. □

Other properties of the Seifert graph pointed out in Murasugi and Przytycki (1991) are the following:

(2.5.4) If the Seifert graph is nonseparable it uniquely determines the underlying link.

(2.5.5) The Seifert graph of a closed m-braid is the block sum of $m-1$ graphs, each consisting of parallel edges. Thus the "natural diagram" of a closed m-braid has exactly m Seifert circles.

An immediate consequence of this is:

(2.5.6) Every link L has at least one diagram D_0 for which the number of Seifert circles $s(D_0)$ equals the braid index $\beta(L)$.

In 1987, S. Yamada proved the following striking result:

(2.5.7) Theorem. For any diagram D of L, $s(D) \geq \beta(L)$.

Combining this with (2.5.6) gives

(2.5.8) $\qquad \beta(L) = \min_{D} s(D)$

where the minimum is taken over all diagrams D representing L.

Example

An extension of Yamada's theorem has recently been obtained by Murasugi and Przytycki using the following combinatorial concept. Define the *cycle index* or *index*, ind(G), of a graph G by:

(2.5.9) $\quad$ ind$(G) = \max_{X \subseteq E(G)} |X| : |X \cap C| < |C|/2$, for all circuits C of G.

The *index* of a link diagram D is the index of the unsigned version of $\Gamma(D)$ and is denoted by ind(D).

This is not the original definition of Murasugi and Przytycki but for bipartite graphs it is equivalent to it by a theorem of Traczyk (1991).

They show that:

(2.5.10) Theorem. For any link diagram D of a link L,

$$\beta(L) \leq s(D) - \text{ind}(D).$$

Since ind(D) is nonnegative, this extends Yamada's theorem, and moreover it is conjectured that

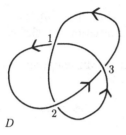 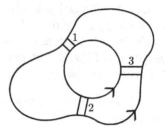

splitting at each crossing

Figure 2.16.

(2.5.11) Conjecture. For an alternating link diagram D of an (alternating) link L,

$$\beta(L) = s(D) - \mathrm{ind}(D).$$

From the complexity point of view this is interesting, for if we suppose that the conjecture is true, then determining the braid index of an alternating link L *given* an alternating diagram representing L, reduces to the problem of finding the index of a planar bipartite graph. Having tried unsuccessfully to develop such an algorithm it did not seem unreasonable to make the conjecture that finding the index of a bipartite planar graph is NP-hard.

Fraenkel and Loebl (1992) now have a proof of this and thus if Conjecture 2.5.11 is true it would mean that:

(2.5.12) Determining the braid index of an alternating link L, even when presented with an alternating diagram, is NP-hard.

However, there is a curiosity here, in that Frank (1989) had previously found a polynomial time algorithm which will solve the following problem

(2.5.13) $\max |X| : |X \cap C| \leq |C|/2$ for all circuits C of G.

Thus this is an interesting example of a situation where a very small change in the constraints has a drastic effect on the difficulty of an optimisation problem.

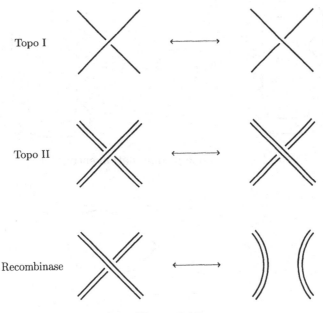

Topo I

Topo II

Recombinase

Figure 2.17.

2.6 Enzyme action

An interesting application of knot theory is what has been described as "the topological approach to enzymology". The object is to unravel the secrets of enzyme action. As there is no known direct method of observing enzymes in action, one approach suggested is to observe the change that the enzyme causes in the topological state of the molecule on which the enzyme is acting.

The DNA molecule can be regarded as a polymer which is long and threadlike, and one of the reasons why knot theory is useful in the study of DNA is that various naturally occurring, enzymes (called topoisomerases) alter the way in which the DNA is embedded in $\mathbf{R}^3$, see for example Wang (1982). These alterations include increasing the "coiling", passing one strand through another by an enzyme induced break in the molecule, breaking strands and rejoining to different ends (recombinant enzymes). Linear DNA is a natural substrate for enzyme action, however this is not of much help to an experimentalist since there can be no interesting observable topology change in an unconstrained linear piece of string. The idea is therefore to get the enzyme to operate on *circular* DNA molecules. This can be done, and then the reaction product should give (by its knot/link structure) a description of the operating enzyme. In other words, the aim is to be able to characterise enzymes by the family of knots which they produce when reacted with unknotted circular substrate. Figure 2.17 shows the local changes which different enzymes cause.

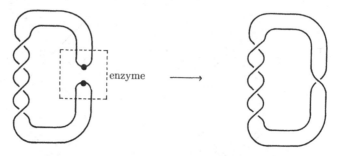

Figure 2.18. A single recombination encounter

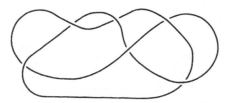

Figure 2.19. The knot 6_2

The first thing to notice is that the action induced by Topo I, namely the switching of a crossing, can lead to the creation of any knot. However, in his very interesting survey Sumners (1987) reports how other enzymes can be characterised by the family of knots (or links) which they produce. For example Phage λ produces only certain types of torus knots.

An extension of this idea was used by Wasserman, Dungan and Cozzarelli (1985) to test a recombination model for a recombinant DNA enzyme, Tn3 resolvase.

Tn3 resolvase is an enzyme which operates on circular DNA. If two resolution sites occur on the same circular duplex DNA molecule they can be oriented either in parallel or in antiparallel, giving rise to links or knots upon recombination. A typical effect of one encounter (or strand exchange) is shown in Figure 2.18.

As a test of the model, the effect was observed of allowing successive encounters of unknotted circular DNA with Tn3.

The experiment achieved four successive strand exchanges on 11 occasions and in each case the knot type produced was the 6_2 knot shown in Figure 2.19. This is exactly the outcome that is predicted by the model.

Since there are 8 distinct 6-crossing links it is argued that the probability of the outcome achieved in their experiment being by chance is $(1/8)^{11} \sim 1.2 \times 10^{-10}$. Since their model explained all the observations the authors concluded it was correct.

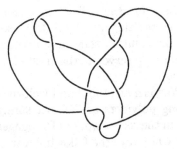 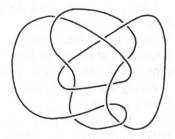

Figure 2.20. The Perko pair

2.7   The number of knots and links

The enzyme classification discussed in the last section leads naturally to the question of how many prime knots and links have crossing number n?

Over time, knot theorists have listed all prime knots with not more than 13 crossings, see Thistlethwaite (1985). However this is an extremely difficult task. For example it was not until 1974 that the Perko pair, shown in Figure 2.20, and until then thought to be different knots with 10 crossings were shown to be the same knot.

Let $k(n)$ and $l(n)$ denote respectively the number of prime knots (links) with crossing number n. By counting the 2-bridge knots of n crossings, Ernst and Sumners (1987) showed that $k(n)$ grows at least exponentially. Their bound gives

(2.7.1) $$\liminf_{n \to \infty} k(n)^{\frac{1}{n}} \geq 2.$$

Clearly $k(n) \leq l(n)$ and in Welsh (1991) it is shown that

(2.7.2) $$4 \leq \liminf_{n \to \infty} l(n)^{\frac{1}{n}} \leq \limsup_{n \to \infty} l(n)^{\frac{1}{n}} \leq 27/2.$$

However the problem of showing that the limit exists is open. It would follow from (2.7.2) and the following:

(2.7.3) Conjecture. Both $k(n)$ and $l(n)$ are supermultiplicative functions, that is,

$$k(m+n) \geq k(m)k(n)$$

$$l(m+n) \geq l(m)l(n).$$

2.8   The topology of polymers

One of the reasons why self-avoiding walks on regular lattices are used as models of linear polymers is that the repulsive interaction between pairs of individual monomers is modelled by forcing the walk to be self-avoiding. In the same way, ring polymers can be modelled by self-avoiding polygons in regular lattices.

Knots are interesting to polymer scientists for the following reason. Highly crystalline polymers are made by crystallising polymers from solutions and as the polymers move through the solution they may become tangled, in other words, knotted. If this tangle is preserved in the crystalline process this may produce defects in the crystal.

A long standing conjecture of Frisch and Wasserman (1968) and Delbruck (1962) can be loosely described as "long ring polymers are almost surely knotted". This was put on a rigorous footing in the recent work of Pippenger (1989) and Sumners and Whittington (1988) who prove the following result.

(2.8.1) Theorem. For even $n \geq 4$, let π_n denote the probability that a random self-avoiding polygon of n steps on the 3 dimensional cubic lattice is unknotted. Then there exists a constant $C < 1$ such that

$$\lim_{n \to \infty} \pi_n^{1/n} = C.$$

The key idea in the proof is the following very beautiful result of Kesten (1963).

A *pattern* is any finite self-avoiding walk, say W. It is a *K-pattern* if there exists some self-avoiding walk on which the pattern W appears three times. Thus if W_1, W_2, W_3 are distinct appearances of W in some longer walk S it must be possible to get out from the end of W_1 to the start of W_2 and back into the beginning of W_3. Kesten's result is:

(2.8.2) Theorem. For any K-pattern W, let $c_n(\epsilon, W)$ be the number of n-step self-avoiding walks on which W appears at most ϵn times. Then there exists $\epsilon > 0$ for which

$$\limsup_{n \to \infty} n^{-1} \log c_n(\epsilon, W) < \kappa,$$

where κ is the connective constant.

In other words, Kesten's theorem says that every K-pattern will appear with positive density on almost all self-avoiding walks.

It is not difficult to extend this to polygons.

(2.8.3) Let $p_n(\epsilon; W)$ be the number of self-avoiding polygons on which the K-pattern W appears at most ϵn times. Then there exists $\epsilon > 0$ such that

$$\limsup_{n \to \infty} n^{-1} \log p_n(\epsilon, W) < \kappa.$$

We now turn to the proof of Theorem 2.8.1.

Proof idea: Let p_n^0 denote the number of unknotted self-avoiding polygons. Then the usual concatenation argument used in self-avoiding walk theory, namely stick the "bottom left" corner of one m-polygon on to the "top right" corner of the other n-polygon to get an $(m + n)$-polygon which is still self-avoiding, shows

$$\lim_{n \to \infty} \left(p_n^0 \right)^{1/n} = e^{\kappa_0} = \beta$$

exists. Now take any fixed knot, such as the trefoil. Embed it in the cubic lattice so tightly (minimally) that the rest of the walk cannot pass through the "centre of this walk" and untie it.

Pippenger (1989) gives an explicit example of such a construction of a walk $W = W(T)$ with the following properties:

(a) it starts at $(0, 0, 0)$ and ends at $(4, 1, 0)$,

(b) apart from $(0, 0, 0)$ and $(4, 1, 0)$ it visits all and only lattice points inside the box

$$B = \{(x, y, z) : \frac{1}{2} \leq x \leq \frac{7}{2}, \quad -\frac{3}{2} \leq y \leq \frac{5}{2}, \quad -\frac{3}{2} \leq z \leq \frac{3}{2}\},$$

(c) it intersects the boundary of B at two points, and does so transversally,

(d) if the portion of the walk lying inside B is closed by a path lying in ∂B the resulting polygon is knotted.

Now using Kesten's theorem, the number of self-avoiding polygons of length n that do not contain at least ϵn occurrences of this walk is at most $(\beta - \epsilon)^n$. Hence the number of self-avoiding polygons is at most $(\beta - \epsilon)^n$, which proves the result. $\qquad\Box$

Early Monte Carlo studies on the probability of random self-avoiding polygons on $\mathbf{Z}^3$ by Vologodskii *et al.* (1974) conclude:

(2.8.4) For small n the probability of being knotted, $1 - \pi_n$, grows linearly with n with

$$1 - \pi_{140} = 0.365 \pm 0.023.$$

This gives an estimate of 0.9968 ± 0.0034 for C.

A major difficulty in this and later Monte Carlo work is how to generate self-avoiding walks (polygons) uniformly at random. We return to this problem in §8.3.

Whittington (1992) reports that recent data are well described by the relation

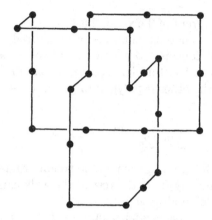

Figure 2.21. Embedding of trefoil in cubic lattice

$$\pi_n \approx 1 - Ce^{-\alpha n}$$

where $\alpha \simeq (5.7 \pm 0.5) \times 10^{-6}$ with C being close to 1 as suggested by the earlier work described above. This is based on walks of size up to $n = 1600$.

In terms of the complexity of the algorithm, for this problem the largest component is not the recognition of knottedness but detecting crossings! But of course this is due to the large number of steps needed to represent a knot in $\mathbf{Z}^3$. For example, the smallest (minimum number of edges) representation of the trefoil in $\mathbf{Z}^3$ is shown in Figure 2.21. It has 24 edges.

It suggests the following:

(2.8.5) Problem: For a knot type K, let $r(K)$ be the minimum number of edges in a representation of K in the cubic lattice. Thus for the unknot U, $r(U) = 4$ and r (Trefoil) $= 24$. Show that finding $r(K)$ from a link diagram representing K is NP-hard.

2.9   Additional notes

An invaluable source for the classical theory of knots is Burde and Zieschang (1985). Apart from this I am also heavily indebted to the articles of Fox (1962), Thistlethwaite (1985) and Kauffman (1988). An early and useful paper on properties of the Tait graph is Yajima and Kinoshita (1957). A clear introduction to the knot group is Crowell and Fox (1977) and the monographs of Stillwell (1980) and Lyndon and Schupp (1977).

Hansen (1989) gives a good introduction to braids; the questions discussed in §2.5 are from Murasugi and Przytycki (1991), Welsh (1992c).

For a recent proof of Yamada's theorem (2.5.7) on the braid index and an improved method of transforming knots into braids see Vogel (1990).

Section 2.6 is based on Sumners (1987a). An interesting treatment in terms of tangles is given by Ernst and Sumners (1990).

For an extension of the ideas of §2.7 to examining entanglements in $\mathbf{Z}^3$ see Soteros, Sumners and Whittington (1992). The representation of the trefoil in Z^3 shown in Figure 2.21 is from Janse van Rensburg and Whittington (1990) and a recent unpublished result of Y. Diao shows it is minimal. For more on the applications of knots in polymer chemistry see Sumners (1987b), and Whittington (1992).

3 Colourings, Flows and Polynomials

3.1 The chromatic polynomial

In 1946 G.D. Birkhoff and D.C. Lewis proposed a quantitative approach to the Four Colour Problem. This concerned properties of the function $P(G;\lambda)$, defined as the number of λ-colourings of a graph G, the hope being that by examining the behaviour of $P(G;\lambda)$, not only for positive integer λ but for arbitrary complex values it might be possible to study the Four Colour Problem by applying methods of real or complex analysis.

For positive integer λ, a λ-*colouring* of a graph G is a mapping of the vertex set $V(G)$ into the set $I_\lambda = \{1, 2, ..., \lambda\}$ such that if $e = (i, j)$ is any edge of G, $\phi(i) \neq \phi(j)$. The members of I_λ are the λ *colours*. We then define $P(G;\lambda)$ to be the number of such mappings ϕ.

Elementary observations are the following.

(3.1.1) If G has n vertices and no edges, then $P(G;\lambda) = \lambda^n$.

(3.1.2) $P(K_n;\lambda) = \lambda^{(n)} = \lambda(\lambda - 1)\ldots(\lambda - n + 1)$.

(3.1.3) If G is the union of two subgraphs H and K such that the intersection $H \cap K$ is the complete graph K_m, then

$$P(G;\lambda) = \frac{P(H;\lambda)P(K;\lambda)}{P(K_m;\lambda)}.$$

(3.1.4) If e is not a loop or an isthmus (= coloop) of G then

$$P(G;\lambda) = P(G'_e;\lambda) - P(G''_e;\lambda)$$

where G'_e and G''_e denote respectively the deletion and contraction of e from G.

Using these, it is easy to give inductive proofs of the following statements.

(3.1.5) If G is a loopless graph with n vertices, then $P(G;\lambda)$ is a polynomial in λ of degree n having integral coefficients.

Accordingly, $P(G;\lambda)$ is now commonly known as the *chromatic polynomial* of G.

Other properties of $P(G;\lambda)$ for any loopless G, which are easily proven, are the following.

(3.1.6) If $P(G; \lambda) = \Sigma a_j \lambda^j$ then a_j is non-zero if and only if $k(G) \leq j \leq |V(G)|$ where $k(G)$ is the number of components.

(3.1.7) $a_j = 1$ when $j = |V(G)|$.

(3.1.8) The non-zero coefficients a_j alternate in sign.

(3.1.9) If G is loopless and $\lambda < 0$, then $P(G; \lambda)$ is non-zero with sign $(-1)^{|V(G)|}$.

A longstanding conjecture about these coefficients, due to Read (1968) is that:

(3.1.10) Conjecture: The sequence of coefficients of the chromatic polynomial is unimodal.

A stronger conjecture due to Hoggar (1974) is:

(3.1.11) Conjecture: The sequence of coefficients is log-concave.

The "golden identity"

There is a huge literature on chromatic polynomials. For planar graphs there are several catalogues in existence, though as we will see, except in very special cases (that is for very special classes of graphs) the computing problem is $\#P$-hard.

I can only skim the surface here, but in order to whet the appetite I highlight a remarkable and mysterious property which was discovered by W.T. Tutte in 1970. It is still not fully understood, nor have its connections with statistical physics been fully worked out.

A *triangulation* is a loopless graph G embedded in the plane in such a way that each face (including the external face) is bordered by exactly 3 edges.

A very curious property of triangulations is summarised in the following statements which combine experimental mathematics/computing with some beautiful theory.

(3.1.12) No example has been found of a triangulation G such that $P(G; \lambda)$ has a real zero in the range $1 < \lambda < 2$ and $\lambda \geq 4$.

(3.1.13) There always seems to be a real zero near $1 + \tau$ where $\tau = \frac{1+\sqrt{5}}{2}$.

(3.1.14) Theorem. If G is a triangulation with n vertices then

$$0 < |P(G; 1 + \tau)| \leq \tau^{5-n}.$$

(3.1.15) The "Golden Identity": If G is a triangulation with n vertices then

$$P(G; 2 + \tau) = (2 + \tau)\tau^{3n-10}P^2(G; 1 + \tau).$$

For more on this fascinating topic see Tutte (1984) and also a recent paper of Woodall (1992) in which he proves.

(3.1.16) Theorem. The chromatic polynomial of a triangulation has no non-integer real zero less than 2.546602 ... .

3.2 The Whitney-Tutte polynomials

In this section we consider 2-variable polynomials which can be regarded as natural extensions of the chromatic polynomial. However, they have a much richer structure, contain many other invariants as specialisations and are of fundamental importance in the later applications to knot theory and statistical physics.

There is some obscurity in the literature concerning the history and nomenclature of these polynomials, see for example Tutte (1984).

First consider the following recursive definition of the function $T(G; x, y)$ of a graph G, and two independent variables x, y.

If G has no edges then $T(G; x, y) = 1$, otherwise for any $e \in E(G)$;

(3.2.1) $T(G; x, y) = T(G'_e; x, y) + T(G''_e; x, y)$, if e is not a loop or isthmus,

(3.2.2) $T(G; x, y) = xT(G'_e; x, y)$ e an isthmus,

(3.2.3) $T(G; x, y) = yT(G''_e; x, y)$ e a loop.

From this, it is easy to show by induction that T is a 2-variable polynomial in x, y, which we call the *Tutte polynomial* of G.

In other words, T may be calculated recursively by choosing the edges in *any* order and repeatedly using (3.2.1-3) to evaluate T. The remarkable fact is that T is well defined in the sense that the resulting polynomial is independent of the order in which the edges are taken. This phenomenon will be a recurrent theme in the development of the various knot polynomials later.

Alternatively, and this is often the easiest way to prove properties of T, we can show that T has the following expansion.

First recall that if $A \subseteq E(G)$, the *rank* of A, $r(A)$ is defined by

(3.2.4)
$$r(A) = |V(G)| - k(A),$$

where $k(A)$ is the number of connected components of the graph $G : A$ having vertex set $V = V(G)$ and edge set A.

It is now straightforward to prove:

(3.2.5) The Tutte polynomial $T(G; x, y)$ can be expressed in the form

$$T(G; x, y) = \sum_{A \subseteq E} (x - 1)^{r(E)-r(A)}(y - 1)^{|A|-r(A)}.$$

This relates T to the *Whitney rank generating function* $R(G; u, v)$ which is a 2-variable polynomial in the variables u, v, and is defined by

(3.2.6) $\qquad R(G; u, v) = \sum_{A \subseteq E} u^{r(E)-r(A)} v^{|A|-r(A)}.$

Using these different interpretations it is not usually too difficult to obtain many of the results which follow. We illustrate the two different methods by proving the following statements.

(3.2.7) For any connected graph G, $T(G; 1, 1)$ counts the number of spanning trees of G.

Proof. Substituting the values $x = 1$, $y = 1$ into the right hand side of (3.2.5), the only terms which count are those for which $r(A) = r(E) = |A|$. These are exactly the spanning trees of G. $\qquad\square$

(3.2.8) For any graph G, the chromatic polynomial $P(G; \lambda)$ is given by

$$P(G; \lambda) = (-1)^{r(E)} \lambda^{k(G)} T(G; 1 - \lambda, 0),$$

where $k(G)$ is the number of connected components of G.

Proof. Verify that with an appropriate weighting $P(G; \lambda)$ satisfies the recursions (3.2.1-3). $\qquad\square$

Far less obvious is the following statement

(3.2.9) For any graph G, $T(G; x, y) = \Sigma t_{ij} x^i y^j$ where the coefficients t_{ij} are non-negative integers.

Tutte (1984) proves this by giving a combinatorial interpretation of the t_{ij}. Comparison with the expansion in terms of powers of $x - 1$ and $y - 1$ given in (3.2.5) makes the property even more surprising.

A convenient way of representing T is as a rectangular array with ij entry t_{ij}.

Example. In the case of K_5 the array is given by

	1	y	y^2	y^3	y^4	y^5	y^6
1	0	6	15	15	10	4	1
x	6	20	15	5	0	0	0
x^2	11	10	0	0	.	.	.
x^3	6	0	.	.	.	.	.
x^4	1	0	.	.	.	.	0

3.3   Tutte Gröthendieck invariants

It is easy and useful to extend these ideas to matroids. The great advantage of matroids is that it enables us to extend the concepts of duality of planar graphs to graphs which are not planar.

Those unfamilar with matroids may think of M as a planar graph, E as its edge set and M^* as its dual plane graph on the same set of edges E, identified in the natural and obvious way.

A *matroid* M is just a pair (E, r) where E is a finite set and r is a submodular *rank function* mapping $2^E \rightarrow \mathbf{Z}$ and satisfying the conditions

(3.3.1) $$0 \leq r(A) \leq |A| \qquad A \subseteq E,$$

(3.3.2) $$A \subseteq B \Rightarrow r(A) \leq r(B),$$

(3.3.3) $$r(A \cup B) + r(A \cap B) \leq r(A) + r(B) \qquad A, B \subseteq E.$$

The edge set of any graph G with its associated rank function as defined by (3.2.4) is a matroid, but this is just a very small subclass of matroids:- known as graphic matroids.

Given $M = (E, r)$ the *dual matroid* $M^* = (E, r^*)$ where r^* is defined by

(3.3.4) $$r^*(E \backslash A) = |E| - r(E) - |A| + r(A).$$

We now just extend the definition of the Tutte polynomial from graphs to matroids by,

(3.3.5) $$T(M; x, y) = \sum_{A \subseteq E(M)} (x-1)^{r(E)-r(A)} (y-1)^{|A|-r(A)}.$$

Much of the theory developed for graphs goes through in this more general setting and there are many other applications as we shall see. For example, routine checking shows that

(3.3.6) $$T(M; x, y) = T(M^*; y, x).$$

In particular, when G is a planar graph and G^* is any plane dual of G, (3.3.6) becomes

(3.3.7) $$T(G; x, y) = T(G^*; y, x).$$

The operations of *deletion* of an edge and *contraction* of an edge in a graph can be described purely in terms of the rank function.

(3.3.8) For the deletion, G'_e (or M'_e), has rank function

$$r'(A) = r(A), \qquad A \subseteq E \backslash e.$$

(3.3.9) For the contraction, G''_e (or M''_e) has rank function r'', given by

$$r''(A) = r(A \cup \{e\}) - r(A), \qquad A \subseteq E \backslash e.$$

These are dual operations in the sense that

$$(M'_e)^* = (M^*)''_e$$

$$(M''_e)^* = (M^*)'_e.$$

A *minor* of M is any submatroid N which can be obtained from M by successively deleting and contracting elements.

The direct sum $M_1 \oplus M_2$ of M_1, M_2 on disjoint sets E_1, E_2 has rank function $r_1 + r_2$ where r_i is the rank function of M_i.

A set X is *independent* if $r(X) = |X|$, it is a *base* if it is a maximal independent subset of E. An easy way to work with the dual matroid M^* is not via the rank function but by the following definition.

(3.3.10) M^* has as its bases all sets of the form $E \backslash B$, where B is a base of M.

We illustrate these concepts with two examples which highlight the much larger class of matroids.

(3.3.11) Example. Uniform matroids. Take E to be any finite set of cardinality say n, and let k be an integer, $0 \le k \le n$. Then U_n^k is the uniform matroid of rank k if its bases are all subsets of cardinality k. Elementary calculation shows

$$T(U_n^k; x, y) = \sum_{r=0}^{k} \binom{n}{r} (x-1)^{k-r} + \sum_{r=k+1}^{n} \binom{n}{r} (y-1)^{r-k}.$$

It is evident from this that $T(U_n^k; x, y) = T(U_n^{n-k}; y, x)$ thus illustrating the duality principle (3.3.7).

(3.3.12) Example. Take any vector space $V(m, q)$ of dimension m over the field $GF(q)$. The natural matroid has as its independent sets just those subsets which are linearly independent. This gives a matroid on *any* subset of vectors; it is a natural vehicle for looking at codes over $GF(q)$ and as we see below, the weight enumerator of a linear code is a special case of the Tutte polynomial. □

We close this section with what we call the "recipe theorem". Its crude interpretation is that whenever a function f on some class of matroids can be shown to satisfy an equation of the form $f(M) = af(M'_e) + b(M''_e)$ for some $e \in E(M)$, then f is essentially an evaluation of the Tutte polynomial. More precisely it says:

(3.3.13) Theorem. Let $\mathcal{C}$ be a class of matroids which is closed under direct sums and the taking of minors and suppose that f is well defined on $\mathcal{C}$ and satisfies

(3.3.14) $\quad f(M) = af(M'_e) + bf(M''_e) \quad$ for $e \in E(M)$ not a loop or coloop,

(3.3.15) $\quad f(M_1 \oplus M_2) = f(M_1)f(M_2)$

then f is given by

$$f(M) = a^{|E|-r(E)}b^{r(E)}T(M; \frac{x_0}{a}, \frac{y_0}{b})$$

where x_0 and y_0 are the values f takes on coloops and loops respectively.

Any invariant f which satisfies (3.3.14)–(3.3.15) is called, a *Tutte-Gröthendieck (TG)-invariant*.

Thus, what we are saying is that any TG-invariant has an interpretation as an evaluation of the Tutte polynomial.

3.4 Reliability theory

Reliability theory deals with the probability of points of a network being connected when individual links or edges are unreliable. Early work in the area was by Moore and Shannon (1956) and now it has a huge literature, see for example Colbourne (1987). Here we shall just give a glimpse of its connections with other problems we have been studying.

Let G be a connected graph in which each edge is independently *open* with probability p and *closed* with probability $q = 1 - p$. The (*all terminal*) *reliability* $R(G; p)$ denotes the probability that in this random model there is a path between each pair of vertices of G. Thus

(3.4.1) $\qquad\qquad R(G; p) = \sum_A p^{|A|}(1 - p)^{|E \setminus A|}$

where the sum is over all subsets A of edges which contain a spanning tree of G, and $E = E(G)$.

It is immediate from this that R is a polynomial in p and a simple conditioning argument shows the following connection with the Tutte polynomial.

(3.4.2) If G is a connected graph and e is not a loop or coloop then

$$R(G; p) = qR(G'_e; p) + pR(G''_e; p),$$

where $q = 1 - p$.

Proof. Condition on the events $\{e \text{ is present}\}$, $\{e \text{ is absent}\}$ and this gives the right hand side. $\qquad\qquad\qquad\qquad\qquad\qquad\qquad\qquad\square$

Using this with the recipe Theorem 3.3.13 it is straightforward to check the following statement.

(3.4.3) Provided G is a connected graph,

$$R(G; p) = q^{|E|-|V|+1} p^{|V|-1} T(G; 1, q^{-1}).$$

3.5 Flows over an Abelian group

Take any graph G, orient its edges arbitrarily. Take any finite Abelian group H and call a mapping $\phi : E(G) \rightarrow H \backslash \{0\}$ a *flow* (or an *H-flow*) if Kirchhoff's laws are obeyed at each vertex of G, the algebra of course being that of the group H.

Note: Standard usage is to describe what we call an H-flow a *nowhere zero H*-flow.

The following statement is somewhat surprising.

(3.5.1) The number of H-flows on G depends only on the order of H and not on its structure.

This is an immediate consequence of the fact that the number of flows is a TG-invariant. To see this, let $F(G; H)$ denote the number of H-flows on G. Then a straightforward counting argument shows that the following is true.

(3.5.2) Provided the edge e is not an isthmus or a loop of G then

$$F(G; H) = F(G_e''; H) - F(G_e'; H).$$

Now it is easy to see that if C, L represents respectively a coloop (= isthmus) and loop then

(3.5.3) $\qquad\qquad F(C; H) = 0 \qquad F(L; H) = o(H) - 1$

where $o(H)$ is the order of H.

Accordingly we can apply the recipe theorem and obtain:

(3.5.4) Theorem: For any graph G and any finite Abelian group H,

$$F(G; H) = (-1)^{|E|-|V|+k(G)} T(G; 0, 1 - o(H)).$$

The observation (3.5.1) is an obvious corollary.

A consequence of this is that we can now speak of G *having a k-flow* to mean that G has a flow over *any* or equivalently *some* Abelian group of order k.

Moreover it follows that there exists a polynomial $F(G; \lambda)$ such that if H is Abelian of order k, then $F(G; H) = F(G; k)$. We call F the *flow polynomial* of G.

The duality relationship (3.3.8) gives:

(**3.5.5**) If G is planar then the flow polynomial of G is essentially the chromatic polynomial of G^*, in the sense that

$$\lambda^{k(G)} F(G; \lambda) = P(G^*; \lambda).$$

A consequence of this and the Four Colour Theorem is that:

(**3.5.6**) Every planar graph having no isthmus has a 4-flow.

Note: the restriction "no isthmus" is the dual of the statement "no loop" for colourings.

What is much more surprising is that the following statement is believed to be true:

(**3.5.7**) **Tutte's 5-Flow Conjecture:** Any graph having no isthmus has a 5-flow.

Indeed it is far from obvious that there is any universal constant k such that graphs without isthmuses have a k-flow. However Jaeger (1976) and Kilpatrick (1975) showed that every such graph had an 8-flow by giving a very elegant argument to show that they had flows over the group $Z_2 \times Z_2 \times Z_2$. More recently Seymour (1981), with a more difficult argument, showed:

(**3.5.8**) **Theorem:** Every graph having no isthmus has a 6-flow.

For more on this and a host of related graph theoretic problems we refer to Jaeger (1988a), here we relate it to a problem of statistical physics.

3.6 Ice models

The simplest ice model concerns the number of ways of orienting the edges of a 4-regular graph in such a way that there are exactly 2 edges oriented inwards and 2 oriented out at each vertex.

This is a more general version of the specific ice problem studied by physicists, when the graph is the $m \times n$ portion of the toroidal square lattice. If Z_{mn} denotes the number of such *ice configurations* the quantity of interest is

$$\lim_{m,n \to \infty} (Z_{mn})^{\frac{1}{mn}} = W.$$

Here it is assumed that the limit exists in the thermodynamic sense.

Pauling (1935) made a rough estimate of W by the following argument.

Assume the orientations of the edges at each vertex are independent configurations. Then if there are N vertices,

$$Z_N \geq 2^{2N} \left(\frac{6}{16} \right)^N = \left(\frac{3}{2} \right)^N.$$

The factor 6/16 comes from the argument that out of a possible 16 ice configurations at each vertex, only 6 are allowable.

The factor 2^{2N} counts all possible assignments of orientations to the $2N$ edges of an N vertex 4-regular square lattice.

In 1967 Lieb showed that:

(3.6.1) For the square lattice $W = \left(\frac{4}{3}\right)^{\frac{3}{2}} = 1.5396007.....$

Lieb's argument is based on transfer matrix arguments and is a beautiful piece of applied mathematics, see Percus (1971).

However it is intriguing that Pauling's approach gave such a close approximation to the true answer. We might also observe that Pauling's argument works equally well for any 4-regular graph.

The connection between ice configurations and nowhere zero flows is the following easily checked statement.

(3.6.2) The number of ice configurations on a 4-regular graph G is given by $F(G; 3)$.

Proof. Given G orient it arbitrarily. Now establish a bijection between flows over the group Z_3 on any oriented version of G with reorientations which correspond exactly to ice configurations. □

3.7 A catalogue of invariants

We now collect together some of the naturally occurring interpretations of the Tutte polynomial. Many of these will not have been encountered yet and all we can do is give a forward reference. Throughout G is a graph, M is a matroid and E will denote $E(G), E(M)$ respectively.

(3.7.1) At (1,1) T counts the number of bases of M (spanning trees in a connected graph).

(3.7.2) At (2,1) T counts the number of independent sets of M, (forests in a graph).

(3.7.3) At (1,2) T counts the number of spanning sets of M, that is sets which contain a base. When G is planar $T(G; 1, 2)$ counts the forests of the dual G^*.

(3.7.4) At (2,0), T counts the number of acyclic orientations of G. Stanley (1973) gives interpretations of T at $(m, 0)$ for general positive integer m, in terms of acyclic orientations.

(3.7.5) Another interpretation at (2,0), and this for a different class of matroids, was discovered by Zaslavsky (1975). This is in terms of counting the number of different arrangements of sets of hyperplanes in n-dimensional Euclidean space.

(3.7.6) $T(G; -1, -1) = (-1)^{|E|}(-2)^{d(B)}$ where B is the bicycle space of G, see Rosenstiehl and Read (1978). When G is planar it also has interpretations in terms of the Arf invariant of the associated knot (link) and the number of components of the link, this is an evaluation of the Jones polynomial of the associated link, see Lickorish (1988).

(3.7.7) The chromatic polynomial $P(G; \lambda)$ is given by

$$P(G; \lambda) = (-1)^{r(E)}\lambda^{k(G)}T(G; 1 - \lambda, 0)$$

where $k(G)$ is the number of connected components.

(3.7.8) The flow polynomial $F(G; \lambda)$ is given by

$$F(G; \lambda) = (-1)^{|E|-r(E)}T(G; 0, 1 - \lambda).$$

(3.7.9) The (all terminal) reliability $R(G : p)$ is given by

$$R(G; p) = q^{|E|-r(E)}p^{r(E)}T(G; 1, 1/q)$$

where $q = 1 - p$.

In each of the above cases, the interesting quantity (on the left hand side) is given (up to an easily determined term) by an evaluation of the Tutte polynomial. We shall use the phrase *"specialises to"* to indicate this. Thus for example, along $y = 0$, T specialises to the chromatic polynomial.

It turns out that the hyperbolae H_α defined by

$$H_\alpha = \{(x, y) : (x - 1)(y - 1) = \alpha\}$$

seem to have a special role in the theory. We note several important specialisations below and they will have further relevance in §6.2.

(3.7.10) Along H_1, $T(G; x, y) = x^{|E|}(x - 1)^{r(E)-|E|}$.

(3.7.11) Along H_2; when G is a graph T specialises to the partition function of the Ising model, see §4.2.

(3.7.12) Along H_q, for general positive integer q, T specialises to the partition function of the Potts model of statistical physics see §4.3.

(3.7.13) Along H_q, when q is a prime power, for a matroid M of vectors over $GF(q)$, T specialises to the weight enumerator of the linear code over $GF(q)$, determined by M. Equation (3.3.6) relating $T(M)$ to $T(M^*)$ gives the MacWilliams identity of coding theory.

(3.7.14) Along H_q for any positive, not necessarily integer, q, T specialises to the partition function of the random cluster model introduced by Fortuin and Kasteleyn (1972). We return to this in §4.5.

(3.7.15) Along the hyperbola $xy = 1$ when G is planar, T specialises to the Jones polynomial of the alternating link or knot associated with G. This connection was first discovered by Thistlethwaite (1987) and will be explained in detail in §5.2.

A host of other more specialised interpretations can be found in the survey of Brylawski and Oxley (1992).

3.8 Additional notes

The origins of chromatic polynomial theory go back at least as far as Birkhoff (1912). More details on the chromatic and Tutte-Whitney polynomials can be found in the books of Aigner (1979), Biggs (1974), Tutte (1984), Welsh (1976). A very detailed survey including many applications is Brylawski and Oxley (1992). However care has to be taken with the parametrisation; there is notable confusion between the dichromate and dichromatic polynomial. Oxley (1992) is a comprehensive up to date account of matroid theory.

Proofs of the golden identity and related theorems can be found in Tutte (1984) or Lovász (1979a).

Crapo (1969) extended the Tutte polynomial to matroids, Brylawski (1972) introduced the Tutte-Gröthendieck approach, the (recipe) Theorem 3.3.14 is from Oxley and Welsh (1979).

Fortuin and Kasteleyn seem to have been the first to recognise the connection between the statistical physics models and the Whitney-Tutte polynomial, but see also Essam (1971), Baxter (1982) and Temperley (1979). The relationship (3.7.13) with linear codes was noticed by Greene (1976). For an extension of some of these concepts to signed objects with particular reference to knots see Kauffman (1989), Jaeger (1992), Murasugi and Przytycki (1991), Traldi (1989), Schwärzler and Welsh (1993), Zaslavsky (1992).

4 Statistical Physics

4.1 Percolation processes

As its name suggests, percolation theory is concerned with flow in random media. Its origin, in 1957 in the work of Broadbent & Hammersley, was as a model for molecules penetrating a porous solid, electrons migrating over an atomic lattice, a solute diffusing through a solvent or disease infecting a community. Here we shall attempt to introduce the main concepts of classical percolation theory and also to relate it with other topics such as the Ising model of ferromagnetism, the reliability problem in random networks, the Ashkin-Teller and Potts models of statistical physics and random clusters.

For illustrative purposes we shall be principally concerned with the two dimensional square lattice L. However the basic ideas apply to any regular lattices in arbitrary dimensions.

Suppose that there is a supply of fluid at the origin and that each edge of L allows fluid to pass along it with probability p, independently for each edge. Let $P_n(p)$ be the probability that at least n vertices of L get wet by the fluid. Thus

$$P_1(p) = 1$$
$$P_2(p) = 1 - (1-p)^4$$

and in theory $P_N(p)$ can be calculated for any integer N. However, the reader will rapidly find it prohibitively time consuming. Obviously

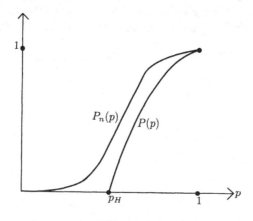

Figure 4.1.

54

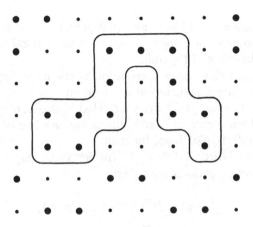

Figure 4.2.

$$P_N(p) \geq P_{N+1}(p)$$

and hence we know that $P(p)$ exists where

(4.1.1) $$P(p) = \lim_{N \to \infty} P_N(p)$$

and it represents the probability that fluid spreads an infinite distance from the origin.

Broadbent & Hammersley (1957) showed that (for a wide class of lattices) there exists a *critical probability* p_H such that

(4.1.2) $$p < p_H \Rightarrow P(p) = 0$$
$$p > p_H \Rightarrow P(p) > 0,$$

and Monte Carlo simulations suggest that for all the well-known lattices the behaviour of $P(p)$ is roughly the same in the qualitative sense as is shown in Fig. 4.1.

Atom or site percolation

Historically, the subject of percolation had statistical mechanics overtones, and in this area 'bond' is usually used to denote an 'edge' of a graph, similarly 'site' or 'atom' denotes a 'vertex'. We shall use these terms interchangeably.

In atom percolation on L instead of each edge of L being randomly blocked with probability $1 - p$ or open with probability p the vertices of L are blocked with probability p or open with probability $q = 1 - p$. Again we are interested in the probability of fluid spreading locally or an infinite distance.

Exactly analogous results hold for atom percolation as for bond percolation, though of course the numerical values of the critical probabilities and percolation probabilities $P(p)$ differ.

It can be argued that atom percolation is the more important, on the grounds that any bond percolation problem on a lattice L can be turned into an atom percolation problem on a related lattice $\tilde{L}$, obtained by letting each edge of L be a vertex in $\tilde{L}$ and joining two vertices of $\tilde{L}$ if and only if the corresponding edges of L are incident.

For any regular latice, if $P^A(p)$, $P^B(p)$ represent respectively the atom and bond percolation probabilities then

(4.1.3) $$P^A(p) \leq P^B(p) \qquad 0 < p < 1,$$

for details see Fisher (1961) and Hammersley (1961).

The cluster problem

Intimately related with percolation theory is the study of the distribution of white and black clusters when the edges (or vertices) of a graph are painted white with probability p and black with probability $q = 1 - p$.

Again we shall concentrate on the edge problem for the square lattice. A black (white) *cluster* is a maximal connected subset of black edges of the lattice where isolated vertices are regarded as clusters. The two main quantities of physical interest are: (a) the average number of white clusters; (b) the average number of vertices in a white cluster.

To be more precise let L_m denote a square section of the square lattice containing m^2 vertices and hence $2(m-1)^2$ edges. If ω denotes a particular black/white painting of L_m then let $c_m(\omega)$ denote the number of white clusters and let its average value over all paintings ω be denoted by $K_m(p)$.

Similarly if we let the distinct clusters under ω be labelled $A_1, ..., A_{c(\omega)}$, we define

$$S_m(p) = \langle \frac{|V(A_1)| + ... + |V(A_{c(\omega)})|}{c_m(\omega)} \rangle$$

where $|V(A_i)|$ denotes the number of vertices in A_i, and $\langle \ \rangle$ denotes the expectation operator over all black-and-white paintings. Thus $S_m(p)$ is the average number of vertices in a white cluster. It is known that

(4.1.4) $$K_m(p) \sim m^2 \lambda(p) \qquad \text{as} \quad m \to \infty$$

where λ is an undetermined function of p, and Temperley & Lieb (1971) have related the problem of enumerating clusters (of at least one white edge) with graph colouring problems, and have shown that as $m \to \infty$,

$$K_m(\frac{1}{2}) \sim 0.098 m^2.$$

It is rather curious that their methods do not seem to extend to $p \neq \frac{1}{2}$.

Roughly speaking the quantities $K_m(p)$ and $S_m(p)$ are reciprocal, and it is straightforward to use the FKG inequality to prove that

$$S_m(p) \geq m^2/K_m(p).$$

For p greater than the critical probability p_H we have positive probability of an infinite white cluster. Hence *a fortiori* as $p \to p_H$ the average size of a cluster tends to ∞. Numerical evidence suggests that as p approaches p_H from below there exists constants C and γ such that as $m \to \infty$, $S_m(p) \to S(p)$ where

$$S(p) \sim C(p_H - p)^{-\gamma}$$

and where moreover γ is an invariant depending only on the dimensionality of the lattice. For a very good discussion of this area see Grimmett (1989) Chapter 7.

The critical probability or probabilities

As stated earlier, p_H, the critical probability, is defined to be the critical value below which there is zero probability that fluid from a source at the origin spreads to infinitely many points. At least two other 'critical probabilities' occur in the literature and there is still confusion about the relationship between them. The first, p_T, is defined to be the critical value of p above which the expected number of points wet by fluid from the origin becomes infinite. Now if there is a positive probability that infinitely many points are wet then *a fortiori* the average number of points wet is infinite. Thus for any lattice,

(4.1.5) $$p_T \leq p_H.$$

Sykes & Essam (1964) in a very ingenious paper, obtained some precise results about a quantity p_E which they call the critical probability but which is defined in terms of singularities of functions giving the mean number of clusters on the lattice. For example, for bond percolation on the square lattice L, they proved that

(4.1.6) $$p_E(L) = \frac{1}{2}$$

and for the triangular lattice T and hexagonal lattice H they showed that

(4.1.7) $$p_E(T) = 2\sin(\pi/18) = 1 - p_E(H).$$

It seems to be extremely difficult to relate p_E with either of the other two critical probabilities p_H and p_T, and physically it does not appear (from its definition at least) to be as natural an object as p_H or p_T. Exact rigorous bounds for p_H and p_T on general lattices seem difficult to obtain. However, for the bond percolation problem on the square lattice, Kesten (1980) showed that $p_T = p_H$ and that this common value was $1/2$. Wierman (1981) extended Kesten's argument and proved a similar result for the hexagonal and triangular lattices thus verifying the earlier result of Essam and Sykes.

For rigorous elegant accounts of the very considerable progress made on percolation problems see the monographs of Kesten (1982) and Grimmett (1989). We return to percolation in §4.5 but close this section by stating two outstanding open problems.

(4.1.8) Problem: Find good bounds or better still, exact values for the critical probabilities of a) site percolation on the square lattice and b) bond or site percolation on the 3 dimensional cubic lattice.

4.2   The Ising model

We now consider a classical model of statistical physics, namely the *Ising model*.

In the general Ising model on a graph or lattice G each vertex i of G is assigned a *spin* σ_i which is either $+1$ (called 'up') or -1 (called 'down'). An assignment of spins to all the vertices of G is called a *configuration* or *state* and is denoted by σ.

In addition each edge $e = (i, j)$ of G has an associated *interaction energy* J_{ij}, which is constant, but may vary from edge to edge. It measures the strength of the interaction between neighbouring pairs of vertices.

For each state $\sigma = (\sigma, ..., \sigma_n)$ define the *Hamiltonian* $H = H(\sigma)$ by

$$(4.2.1) \qquad H(\sigma) = -\sum_{(ij)} J_{ij}\sigma_i\sigma_j - \sum_i M\sigma_i,$$

where M is the energy due to the external field.

The Hamiltonian $H(\sigma)$ measures the energy of the state σ.

In a ferromagnet the J_{ij} are positive; this means, that a configuration of spins in which nearest neighbour pairs have parallel spins ($\sigma_i = \sigma_j$) has a lower energy than a non-magnetised state in which spins are arbitrary.

The external field M has an effect of aligning spins with the direction of the field, thus again favouring states of low energy.

The *partition function* $Z = Z(G, \beta, J, M)$ is defined by

$$(4.2.2) \qquad Z = \sum_\sigma e^{-\beta H(\sigma)},$$

where the sum is over all possible spin configurations σ with $\sigma_i \in \{-1, 1\}$, and $\beta = 1/kT$ is a parameter determined by the temperature T (in absolute degrees) and where k is Boltzmann's constant. The importance of Z is that it is assumed that the probability of finding the system in a state or configuration σ, is given by

$$(4.2.3) \qquad Pr(\sigma) = e^{-\beta H(\sigma)}/Z.$$

Thus we see that

(i) High temperature $\Rightarrow$ low value of β $\Rightarrow$ probability distribution of states becomes more flat.

(ii) Low temperature $\Rightarrow$ high β $\Rightarrow$ greater probability to low energy states.

Entropy and free energy

For any finite probability distribution $(p_1, ..., p_N)$, the *entropy* $h(p_1, ..., p_N)$ is defined by

$$h = -\sum_k p_k \log_2 p_k$$

and is a measure of the uncertainty (or randomness) in the system. Thus, in the case of the stochastic Ising model we have:

$$h(\beta, G) = -\sum_\sigma Pr(\sigma) \log_2 Pr(\sigma),$$

which after substituting from (4.2.3) gives

$$h(\beta, G) = \left[\frac{1}{Z}(\beta \log_2 e) \sum_\sigma H(\sigma) e^{-\beta H(\sigma)}\right] + \log_2 Z.$$

But

$$\frac{\partial}{\partial \beta} \log Z = \frac{1}{Z} \frac{\partial Z}{\partial \beta} = -\sum_\sigma \frac{H(\sigma)}{Z} e^{-\beta H(\sigma)}.$$

Thus

(4.2.4) $$h(\beta, G) = -(\beta \log_2 e) \frac{\partial}{\partial \beta} \log Z + \log_2 Z.$$

The quantity

$$U = -\frac{\partial}{\partial \beta} \log Z$$

is called the *internal energy*, and the *free energy* F is defined to be $\log Z$. Thus the entropy h can be represented as a sum

$$h(\beta, G) = (\beta \log_2 e)U(G) + (\log_2 e)F(G).$$

The main problem of the Ising model on a given lattice is to find a closed expression for

(4.2.5) $$\lim_{n \to \infty} F(G_n)/n = \lim_{n \to \infty} n^{-1} \log Z(G_n)$$

where G_n is a sequence of graphs approaching (in some reasonable sense) the infinite lattice graph. There is no guarantee that the limit in (4.2.5) is well defined or even when well defined will exist. On the *assumption* that it does, it is called the *free energy per lattice site*.

Another interesting quantity is the *pair* or *two-point correlation* function

$$\langle \sigma_i, \sigma_j \rangle = \left[\sum_\sigma \sigma_i \sigma_j \exp(-\beta H(\sigma))\right] /Z.$$

This is a natural measure of disorder in the lattice and as we shall see in §4.5 is closely related to percolatory behaviour in the random cluster model.

4.3   Combinatorial interpretations

Turn now to a combinatorial analysis of the Ising model on a finite graph $G = \langle V, E \rangle$. We will assume first that:

(i) there is no external magnetic field, thus $M = 0$;

(ii) the interaction energy is constant for each edge and equal to J.

In this case the partition function Z reduces to

$$(4.3.1) \qquad Z = Z(G, \beta, J) = \sum_{\sigma} e^{-\beta H(\sigma)}$$

where

$$(4.3.2) \qquad H(\sigma) = -\sum_{ij \in E} J\sigma_i \sigma_j.$$

Now consider a given spin configuration σ, it can be regarded as a partition of the vertex set V into V^+, the set of i for which $\sigma_i = +1$, and V^-, the set of i for which $\sigma_i = -1$.

Thus if $E^+(E^-)$ denote respectively the sets of edges having both endpoints in $V^+(V^-)$, then the energy of the state σ is given by

$$
\begin{aligned}
H(\sigma) &= -J\left(\left|E^+\right| + \left|E^-\right|\right) + J\left(\left|\delta(V^+, V^-)\right|\right) \\
&= +J\left(-|E| + 2\left|\delta(V^+, V^-)\right|\right),
\end{aligned}
$$

where $\delta(V^+, V^-)$ is the set of edges of the bipartition of G induced by V^+, V^-.

A *bipartition* of a graph $G = \langle V, E \rangle$ is a subgraph $B = \langle V, E' \rangle$ which is bipartite.

Each bipartition B of G will correspond to exactly two spin configurations (depending on which side is $+ve$ or $-ve$). However the contribution to Z will be the same in both cases and given by

$$H(\sigma) = H(B_\sigma) = -J\,|E(G)| + 2J\,|E(B)|.$$

Thus we see

$$
\begin{aligned}
Z &= \sum_{\sigma} e^{-\beta H(\sigma)} \\
&= 2\sum_{B} e^{+\beta J|E(G)| - 2\beta J|E(B)|}
\end{aligned}
$$

where the sum is over all bipartitions B. Therefore

$$Z(G) = 2e^{\beta J|E(G)|} \sum_{B} e^{-2\beta J|E(B)|}.$$

Thus if b_m is the number of bipartitions of G with m edges, we can write $\theta = e^{-2\beta J}$ and then

$$(4.3.3) \qquad Z(G) = 2\theta^{-|E(G)|/2} \sum_{m=0}^{|E(G)|} b_m \theta^m.$$

An Eulerian expansion for Z

It was observed by van der Waerden (1941) that in the absence of an external magnetic field the partition function has an expansion in terms of *Eulerian* subgraphs, that is subgraphs in which every vertex degree is even.

To see this let G be finite and take $K = -\beta J$ so that we can write $Z_G(K)$ in the form

$$
\begin{aligned}
Z_G(K) &= \sum_\sigma \exp(K \sum_{ij \in E} \sigma_i \sigma_j) \\
&= \sum_\sigma \prod_{ij \in E} \exp(K \sigma_i \sigma_j) \\
&= \sum_\sigma \prod_{ij \in E} \cosh(K \sigma_i \sigma_j)[1 + \tanh(K \sigma_i \sigma_j)] \\
&= \sum_\sigma \prod_{ij \in E} \cosh(K)[1 + \sigma_i \sigma_j \tanh(K)]
\end{aligned}
$$

using the fact that $\sigma_i, \sigma_j \in \{-1, 1\}$. Thus

$$
\begin{aligned}
Z &= (\cosh(K))^{|E|} \sum_\sigma \sum_{A \subseteq E} \prod_{ij \in A} (\sigma_i \sigma_j \tanh(K)) \\
&= (\cosh(K))^{|E|} \sum_{A \subseteq E} (\tanh(K))^{|A|} \sum_\sigma \prod_{ij \in A} \sigma_i \sigma_j.
\end{aligned}
$$

Now if $\mathcal{C}$ is the set of Eulerian subgraphs of G; then we see

$$
A \notin \mathcal{C} \implies \sum_\sigma \prod_{ij \in A} \sigma_i \sigma_j = 0;
$$

(4.3.4)
$$
A \in \mathcal{C} \implies \prod_{ij \in A} \sigma_i \sigma_j = 1.
$$

Thus

(4.3.5)
$$
\begin{aligned}
Z(G) &= (\cosh(K))^{|E|} \sum_{A \in \mathcal{C}} (\tanh(K))^{|A|} 2^{|V|} \\
&= (\cosh(K))^{|E|} 2^{|V|} C_G(\tanh(K))
\end{aligned}
$$

where $C_G(x)$ is the generating function of the number of Eulerian subgraphs of G.

The Ising model and the Tutte polynomial

It is not difficult to show that in the absence of an external magnetic field, and with $J_e = J$ for all edges e, then whenever e is not a loop or coloop of G,

(4.3.6)
$$
Z(G) = e^{\beta J} Z(G'_e) + 2 \sinh(\beta J) Z(G''_e).
$$

Also consider the graphs C consisting of a single edge and L consisting of a single loop. Then

$$Z(C) = 2e^{\beta J} + 2e^{-\beta J} = 4\cosh(\beta J)$$
$$Z(L) = 2e^{\beta J}.$$

Thus, applying the recipe theorem (3.3.14) we get the result

(4.3.7) $Z(G) = (2e^{-\beta J})^{|E|-r(E)}(4\sinh\beta J)^{r(E)}T(G; \coth\beta J, e^{2\beta J}).$

Thus we have obtained three distinct expansions for Z.

Dealing with an external field

So far we have been considering the restricted case where the external field is zero. Here we show how determining the partition function of the Ising problem with a non-zero field on G, can be transformed into a zero field problem on an augmented graph $\hat{G}$.

The construction of $\hat{G}$ from G is easy, just adjoin an additional vertex v_0 to G and let additional edges v_0v_i $(1 \le i \le n)$ have constant interaction energy X. Then the partition function $\hat{Z}$ of $\hat{G}$ with no external field but with interaction energies inherited from G can be written as

(4.3.8) $\hat{Z} = \sum_\sigma e^{-\beta[-\Sigma J_{ij}\sigma_i\sigma_j]}(e^{-\beta\Sigma X\sigma_i} + e^{+\beta\Sigma X\sigma_i})$

where the sum is over all states σ on $V(G)$ and the additional multiplying factor comes from the spins ± 1 on the new vertex v_0.

Now consider the expression (4.3.6) for $2Z(G)$ obtained by pairing together the states

$$\sigma = (\sigma_1, ..., \sigma_n) \quad \text{and} \quad -\sigma = (-\sigma_1, -\sigma_1, ..., -\sigma_n).$$

This gives

(4.3.9) $2Z(G) = \sum_\sigma e^{-\beta[-\Sigma J_{ij}\sigma_i\sigma_j]}(e^{-\beta M\Sigma\sigma_i} + e^{\beta M\Sigma\sigma_i})$

Comparing (4.3.6) and (4.3.7) shows that by taking $X = M$ gives equality and thus we have proved our original assertion.

4.4   The Ashkin-Teller-Potts model

There is a straightforward generalisation of the Ising model in which each atom can be in Q different states $(Q \ge 2)$. In this model introduced by Ashkin and Teller (1943) and Potts (1952) the energy between two interacting spins is taken to be zero if the spins are the same and equal to a constant if they are different. If we now denote this constant associated with an edge (ij) by K_{ij} then in state σ, the *Hamiltonian* $H(\sigma)$ is defined by

$$H(\sigma) = \sum_{(ij)} K_{ij}(1 - \delta(\sigma_i, \sigma_j))$$

where δ is the usual Kronecker delta function.

The *partition function* Z is again defined by

(4.4.1) $$Z = \sum_{\sigma} \exp[-H(\sigma)]$$

where the sum is over all possible spins σ. Thus in a lattice of n vertices the sum in (4.4.1) is over Q^n possible states.

Note: We are assuming a zero external magnetic field. Otherwise Z is modified as in the case of the Ising model.

Suppose now that we partition the edge set E into $E^+ \cup E^-$ where E^+ (E^-) respectively denotes the sets of edges whose endpoints are the same (different) under a given state σ.

Then the contribution of σ to the Hamiltonian will be $2K(E^-)$ where

$$K(E^-) = \sum_{ij:\sigma_i \neq \sigma_j} K_{ij}.$$

In the particular case where K_{ij} is constant over the edges, and comparing with the corresponding expression for the Ising model, we see that if we take $K_{ij} = -2J_{ij}$ then

$$Z_{\text{Potts}} = e^{-\Sigma J_{ij}} Z_{\text{Ising}}$$

where in both cases we assume we have zero external field.

For the rest of this discussion we will assume $J_{ij} = J$ is constant, so that we can write $K = 2\beta J$ and then

(4.4.2)
$$\begin{aligned}
Z(G)_{\text{Potts}} &= \sum_{\sigma} \exp(-H(\sigma)) \\
&= \sum_{\sigma} e^{-K|E^-(\sigma)|}.
\end{aligned}$$

As we now see this is just another specialisation of the Tutte polynomial.

The monochrome polynomial

Let $b_i(\lambda)$ be the number of λ-colourings of the vertex set V of a graph G, in which there are i *monochromatic* or *bad* edges, that is they have endpoints of the same colour.

Consider the generating function

$$B(G; \lambda, s) = \sum_{i=0}^{|E|} s^i b_i(\lambda).$$

Clearly $b_0(\lambda)$ is the chromatic polynomial of G and like $P_G(\lambda)$ we see that the following relationships hold.

(4.4.3) If G is connected then provided e is not a loop or coloop,

$$B(G; \lambda, s) = B(G'_e; \lambda, s) + (s-1)B(G''_e; \lambda, s).$$

(4.4.4) $B(G; \lambda, s) = sB(G'_e)$ if e is a loop.

(4.4.5) $B(G; \lambda, s) = (s + \lambda - 1)B(G''_e)$ if e is a coloop.

Combining these, we get by using the recipe theorem (3.3.14)

(4.4.6) $B(G; \lambda, s) = \lambda(s-1)^{|V|-1}T(G; \frac{s+\lambda-1}{s-1}, s)$.

Consider now the relation with the Potts model. From (4.4.2) we can write

$$
\begin{aligned}
Z_{\text{Potts}}(G) &= \sum_\sigma e^{-K|E^-(\sigma)|} \\
&= e^{-K|E(G)|}\sum_\sigma e^{K|E^+(\sigma)|} \\
&= e^{-K|E|}\sum_{Q\text{-colourings}} b_j(Q)(e^K)^j \\
&= e^{-K|E|}B(G; Q, e^K).
\end{aligned}
$$

Then using the relationship (4.4.6) we get,

(4.4.7) $Z_{\text{Potts}}(G) = Q(e^K - 1)^{|V|-1}e^{-K|E|}T\left(G; \dfrac{e^K + Q - 1}{e^K - 1}, e^K\right).$

It is not difficult to verify that $T(G; x, y)$ can be recovered from the monochrome polynomial and therefore from the Potts partition function by using the formula

(4.4.8) $T(G; x, y) = \dfrac{1}{(y-1)^{|V|}(x-1)}B(G; (x-1)(y-1), y).$

Note that this formula highlights again the special nature of the hyperbolae $H_\alpha \equiv (x-1)(y-1) = \alpha$, in this theory.

4.5 The random cluster model

The general random cluster model on a finite graph G introduced by Fortuin and Kasteleyn (1972) is a correlated bond percolation model on the edge set E of G defined by the probability distribution,

(4.5.1) $\mu(A) = Z^{-1}\left(\displaystyle\prod_{e \in A} p_e\right)\left(\displaystyle\prod_{e \notin A}(1 - p_e)\right)Q^{k(A)}$ $(A \subseteq E),$

where $k(A)$ is the number of connected components (including isolated vertices) of the subgraph $G : A = (V, A)$, p_e $(0 \le p_e \le 1)$ are parameters associated with each edge of G, $Q \ge 0$ is a parameter of the model, and Z is the normalising constant introduced so that

$$\sum_{A \subseteq E} \mu(A) = 1.$$

We will sometimes use $\omega(G)$ to denote the random configuration produced by μ, and P_μ to denote the associated probability distribution.

Thus, in particular, $\mu(A) = P_\mu\{\omega(G) = A\}$. When $Q = 1$, μ is what Fortuin and Kasteleyn call a *percolation model* and when each of the p_e are made equal, say to p, then $\mu(A)$ is clearly seen to be the probability that the set of *open* edges is A in bond percolation.

For an account of the many different interpretations of the random cluster model we refer to the original paper of Fortuin and Kasteleyn or to Sokal (1989).

Here we shall be concentrating on the percolation problem when each of the p_e are equal, to say p, and henceforth this will be assumed.

Thus we will be concerned with a two parameter family of probability measures

$$\mu = \mu(p, Q) \quad \text{where} \quad 0 \le p \le 1 \quad \text{and} \quad Q > 0$$

defined on the edge set of the finite graph $G = (V, E)$ by

$$\mu(A) = p^{|A|} q^{|E \setminus A|} Q^{k(A)} / Z$$

where Z is the appropriate normalising constant, and $q = 1 - p$.

The reason for studying percolation in the random cluster model is its relation with phase transitions via the two-point correlation function. This was pointed out first by Fortuin and Kasteleyn and given further prominence recently by Edwards and Sokal (1988) in connection with the Swendsen-Wang algorithm (1987) for simulating the Potts model. We describe briefly the connection.

Let Q be a positive integer and consider the Q-state Potts model on G.

The probability of finding the system in the state σ is given by the Gibbs distribution

$$Pr(\sigma) = e^{-H(\sigma)} / Z.$$

The key result is the following:

(4.5.2) Theorem. For any pair of sites (vertices) i, j, and positive integer Q, the probability that σ_i equals σ_j in the Q-state Potts model is given by

$$\frac{1}{Q} + \frac{(Q-1)}{Q} P_\mu\{i \rightsquigarrow j\}$$

where P_μ is the random cluster measure on G given by taking $p_e = 1 - \exp(-J_{ij})$ for each edge $e = (ij)$, and $\{i \leadsto j\}$ is the event that under μ there is an open path from i to j.

The attractive interpretation of this is that the expression on the right hand side can be regarded as being made up of two components.

The first term, $1/Q$, is just the probability that under a purely random Q-colouring of the vertices of G, i and j are the same colour. The second term measures the probability of long range interaction. Thus we interpret the above as expressing an equivalence between long range spin correlations and long range percolatory behaviour.

Phase transition (in an infinite system) occurs at the onset of an infinite cluster in the random cluster model and corresponds to the spins on the vertices of the Potts model having a long range two-point correlation.

Thus the random cluster model can be regarded as the analytic continuation of the Potts model to non integer Q.

In order to be able to calculate or even simulate the Gibbs state probabilities it seems to be necessary to know (or be able to approximate) the partition function Z. In the case of ordinary percolation, $Q = 1$, and $Z = 1$, but in general, determining Z is equivalent to determining the Tutte polynomial, as it is not difficult to show.

(4.5.3) For any finite graph G and subset A of $E(G)$, the Gibbs probability μ is given by

$$\mu(A) = \frac{\left(\frac{p}{q}\right)^{|A|} Q^{-r(A)}}{\left(\frac{p}{Qq}\right)^{r(E)} T(G; 1 + \frac{Qq}{p}, \frac{1}{q})},$$

where T is the Tutte polynomial of G, where $q = 1 - p$, and where r is given by $k(A) = |V(G)| - r(A)$,

A first consequence of this is that, as we see in §6.5, determining the Gibbs probability μ is an intractable problem for most Q and most graphs.

An obvious quantity of interest is the probability that a particular set is open. We call this the *distribution function*, denote it by λ, and note that it is given by

$$\lambda(A) = \sum_{X: X \supseteq A} \mu(X).$$

The sort of questions we need to be able to answer are, how does λ vary with p and Q and how difficult is it to calculate λ?

A very useful result which follows easily from an extension of the FKG inequality by Holley (1974) is the following.

(4.5.5) Proposition. Provided $1 \leq Q_1 \leq Q_2$, for any fixed p, $0 \leq p \leq 1$ and any nondecreasing function $f : 2^E \to \mathbf{R}$,

$$\langle f \rangle_{\mu_1} \geq \langle f \rangle_{\mu_2}$$

where μ_1 and μ_2 are the random cluster measures induced by p and Q_1, Q_2 respectively.

A special case of this gives

(4.5.6) Corollary. For fixed p, the distribution function λ is a monotone nonincreasing function of Q, for $Q \geq 1$.

A question which seems difficult is the following.

(4.5.7) Problem. How does λ vary with Q when $0 < Q < 1$?

4.6 Percolation in the random cluster model

Consider a random cluster model $\mu = \mu(p, Q)$ on E the edge set E of a planar graph G and let G^* be the dual plane graph with edge set also E identified in the natural and obvious way.

Now define the *dual measure* $\hat{\mu}$ of $\mu = \mu(p, Q)$ to be the random cluster measure $\hat{\mu}(\hat{p}, \hat{Q})$ where

$$\hat{p} = \frac{qQ}{p + qQ}, \quad \hat{Q} = Q.$$

Thus

$$\hat{\mu}(A) \propto \left(\frac{qQ}{p} \right)^{|A|} Q^{-r(A)}.$$

(4.6.1) Proposition. For any plane graph G and random cluster measure μ

$$P_\mu\{\omega(G) = A\} = P_{\hat{\mu}}\{\omega(G^*) = E \backslash A\}.$$

(4.6.2) Corollary. If G, G^* are dual planar graphs, $\hat{\mu}$ on G^* produces white configurations with exactly the same probability distribution as μ produces black configurations on G.

Proof.

$$P_{\hat{\mu}}\{\omega(G) = E \backslash A\} \propto \left(\frac{Qq}{p} \right)^{|E \backslash A|} Q^{-r^*(E \backslash A)}$$

$$\propto \left(\frac{p}{q} \right)^{|A|} Q^{-r(A)}$$

by using the duality formula (3.3.4) for rank functions. □

We now turn to the specific case of the square lattice. We adopt the terminology of ordinary $(Q = 1)$ percolation as much as possible and in particular follow the notation of Grimmett (1989).

Let Λ_n denote the box on the square lattice having corners $(\pm n, \pm n)$. Let p, Q be fixed and let $\mu_m = \mu_m(p, Q)$ be the sequence of random cluster measures induced by Λ_m, as m runs through the positive integers.

The events in which we have a particular interest are of type $\{0 \rightsquigarrow \partial_n\}$ denoting the event that there is an open path from 0 to ∂_n, the boundary of the box Λ_n.

(4.6.3)   For $Q \geq 1$ and $m \geq n$,

$$\mu_{m+1}\{0 \rightsquigarrow \partial_n\} \geq \mu_m\{0 \rightsquigarrow \partial_n\}.$$

This is just a special case of the following:

(4.6.4) Proposition. Let G be a finite graph and let H be a subgraph of G on the same vertex set. If μ_G and μ_H denote the random cluster measures induced by G, H respectively for any fixed p and $Q \geq 1$, then for any monotone nondecreasing f on the edge set of G, if the value of f is determined by the state of the edges of H, then

$$\langle f \rangle_{\mu_H} \leq \langle f \rangle_{\mu_G}.$$

Since the quantities in (4.6.1) are probabilities and thus bounded, we can therefore define

$$\theta_n(p, Q) = \lim_{m \to \infty} \mu_m\{0 \rightsquigarrow \partial_n\}.$$

Now for $m > n$, it is trivial that

$$\mu_m\{0 \rightsquigarrow \partial_n\} \leq \mu_m\{0 \rightsquigarrow \partial_{n-1}\}.$$

Consequently

$$\theta_n(p, Q) \leq \theta_{n-1}(p, Q)$$

and we define

$$\theta(p, Q) = \lim_{n \to \infty} \theta_n(p, Q)$$

to be the *percolation probability* of the model.

Note that when $Q = 1$, $\theta(p, Q)$ is essentially the same quantity as $P(p)$ defined in §4.1. Accordingly, for $Q \geq 1$, we can define the *critical probability* $p_c(Q)$ by

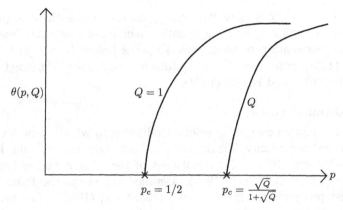

Figure 4.3.

$$p_c(Q) = \inf p : \theta(p, Q) > 0.$$

We know from Theorem (4.6.2) that:

(4.6.5)   For $Q \geq 1$, $p_c(Q)$ is monotone nondecreasing in Q.

In Welsh (1992c) it is shown that the following is true.

(4.6.6) For $Q \geq 1$, the critical probability $p_c(Q)$ satisfies

$$\frac{\sqrt{Q}}{1 + \sqrt{Q}} \leq p_c(Q).$$

It would be natural to expect that for any $Q \geq 1$, the quantities $\theta(p, Q)$ looked roughly like $P(p)$. In other words, the situation is as sketched in Figure 4.3.

I believe that the following Q-extension of Kesten's Theorem is true.

(4.6.7) **Conjecture.** For $Q \geq 1$, the critical probability $p_c(Q)$ equals $\sqrt{Q}/(1 + \sqrt{Q})$.

The conjecture is certainly true when $Q = 1$ by Kesten's theorem that the critical probability of the square lattice is $\frac{1}{2}$. It is also true when $Q = 2$ because using the relation $p = 1 - e^{-J}$, when $Q = 2$, this corresponds to a critical value of $\sinh^{-1} 1 = 0.88137$ for the critical exponent J, agreeing with the Onsager solution to the Ising model, see Hammersley and Mazzarino (1983).

For integer $Q \geq 3$ the critical value of $p_c(Q)$ given by the conjecture agrees with the critical points of the Potts model located by singularity based arguments see for example Baxter (1982) or Hintermann, Kunz and

Wu (1978). However it does not appear easy to make these arguments rigorous in this context, and the situation seems not dissimilar from that in ordinary percolation when it took 16 years before Kesten (1980) and Wierman (1981) were able to give rigorous justifications of the exact values obtained by Sykes and Essam (1964).

4.7 Additional notes

Section 4.1 is just an extremely brief introduction to what is now a vibrant area of applied probability. There exist excellent texts, in particular Kesten (1982), Grimmett (1989). The treatment of the Ising model is based on Thompson (1972) and Cipra (1987). Detailed surveys of the Potts model and related problems are given by Wu (1982) and (1984). The inversion formula (4.4.8) is contained in Tutte (1984), though not in this form. Fortuin and Kasteleyn in their original paper pointed out the connection between their model and the Tutte polynomial. A detailed discussion of the random cluster model is given by Aizenmann, Chayes, Chayes and Newman (1988) and Bezuidenhout, Grimmett and Kesten (1992). The treatment given here follows Welsh (1992c). A good account of Holley's theorem and the FKG inequality and their relation to problems of statistical physics is given in Preston (1974). A remarkable paper by Laanit *et al.* (1991) shows that Conjecture 4.6.7 is true for sufficiently large Q, while a treatment of the general cluster distribution problem for a wide range of Gibbs probability distributions is given by Gandolfi, Keane and Newman (1992).

5 Link Polynomials and the Tait Conjectures

5.1 The Alexander polynomial

One classical invariant of knots which we have not yet discussed is the Alexander polynomial, introduced by J.W. Alexander in 1928, and much used as a separator of knots.

The Alexander polynomial is an invariant of *oriented* knots and was traditionally defined as an evaluation of a determinant of a matrix determined by the knot diagram. However, in 1969 J. Conway introduced a polynomial invariant of oriented links which turned out to be essentially the same as the Alexander polynomial. This leads to a recursive definition of the Alexander polynomial as we now show. To each oriented link L is associated a

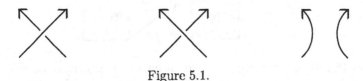

Figure 5.1.

polynomial $\nabla_L(z)$ belonging to the polynomial ring $Z[z]$. Equivalent knots and links receive the same polynomial which satisfies the following recursive rules.

(5.1.1) If $L \sim U$, the unknot, then $\nabla_L(z) = 1$.

(5.1.2) If links differ at the one crossing as shown in Figure 5.1 but are otherwise the same, then

$$\nabla_L(\nwarrow\!\!\nearrow) = \nabla_L(\nearrow\!\!\nwarrow) + z\nabla_L(\;)(\;).$$

We call $\nabla_L(z)$ the *Conway polynomial* of L.

The relationship shown in Figure 5.1 is an example of what is known as a *skein relation*. These are recursive relations relating invariants of links which are identical except in local regions of the link diagram.

The *Alexander polynomial* of L written Δ_L is defined in terms of the Conway polynomial by

$$(5.1.3) \qquad \Delta_L(t) = \nabla_L(\sqrt{t} - \frac{1}{\sqrt{t}}).$$

The Alexander polynomial is well understood, in contrast to later polynomials we will consider.

First we show that if L is a *split link* or is *splittable* in the sense that it consists of 2 non empty parts which can be separated by a 2-sphere then its Conway polynomial is zero. The argument follows Kauffman (1987b).

(5.1.4) If L is splittable then $\nabla_L = 0$.

Proof. If L is splittable with say $L = L_1 \cup L_2$, then it can be oriented as shown.

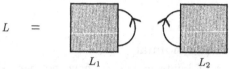

$$L \quad = \qquad\qquad\qquad L_1 \qquad\qquad\qquad L_2$$

Now note that if M and N are constructed from L as shown then M is isotopic to N and thus $\nabla_M = \nabla_N$. By condition (5.1.2) ∇_L must equal zero.

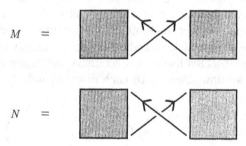

$$M \quad =$$

$$N \quad =$$

$\square$

Using (5.1.4) and the recursive definitions it is algorithmically straightforward though time consuming (exponentially slow) to obtain the Conway polynomial of a link by a succession of switches as we now illustrate.

Example. Consider the trefoil T

which by (5.1.2) yields $\nabla_T = \nabla_L + z\nabla_M$ where

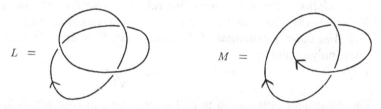

$$L \quad = \qquad\qquad\qquad\qquad\qquad\qquad M \quad =$$

Figure 5.2.

L is the unknot so $\nabla_L = 1$. Using (5.1.2) again we have $\nabla_M = \nabla_{M_1} + z\nabla_{M_2}$

But M_1 is split, so $\nabla_{M_1} = 0$ and $\nabla_{M_2} = \nabla_U = 1$,
   Combining these relations gives

$$\nabla_T(z) = 1 + z^2.$$

To obtain the Alexander polynomial of the trefoil use (5.1.3) to get

$$\begin{aligned}
\Delta_T(t) &= \nabla_T(\sqrt{t} - 1/\sqrt{t}) \\
&= 1 + t - 2 + \frac{1}{t}
\end{aligned}$$

which on normalisation (as is usual) gives

$$\Delta_T(t) = 1 - t + t^2.$$

$\square$

Proof of invariance
In order to show that a given polynomial is in fact a knot/link invariant, it is necessary and sufficient to show that the invariant in question is unchanged under each of the three Reidemeister moves.

   In the case of oriented knots/links, we allow any consistent orientation of the link diagrams, for example, the oriented version of the third move would allow the move shown in Figure 5.2 We will illustrate this in detail in the next section.

   First note that it is not difficult to find nontrivial knots which have the same Alexander polynomial as the unknot - one such example is the pretzel knot (- 3,5,7). Figure 5.3 shows the general pretzel link which is parametrised by $(c_1, c_2, ..., c_n)$ where c_i is the number of crossings in the ith tassle, where if c_i is negative the crossings are in the opposite sense.

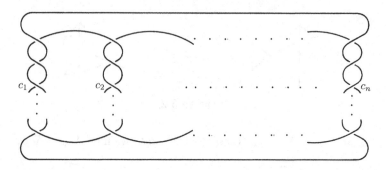

Figure 5.3.

It is easy to prove that for any knot, $\Delta(1) = \pm 1$ and then the usual convention is that Δ is normalised so that it is of the form

$$\Delta(t) = a_0 + a_1 t + \ldots + a_n t^n$$

with $\Delta(1) = 1$.

Well known properties of Δ are the following:

(5.1.5) its coefficients form a palindromic sequence, that is $\Delta(t) = t^{deg\Delta}\Delta(t^{-1})$;

(5.1.6) it has even degree;

(5.1.7) $|\Delta(-1)|$, is called the *determinant* of the knot and is an odd integer, it was one of the first invariants studied;

(5.1.8) if K is an alternating knot, Δ_K is an alternating polynomial and its degree is twice the genus of the knot.

There is also a *conjecture* that for

$$1 \le i \le \frac{1}{2}deg\Delta, \quad |a_{i-1}| \le |a_i|.$$

It is interesting to compare this with the corresponding unimodality conjecture for chromatic polynomials see (3.1.10).

A rather striking result, for a proof see Burde and Zieschang (1985), is the following theorem.

(5.1.9) Theorem. If $p(t)$ is any polynomial satisfying $p(1) = 1$ and $p(1/t) = p(t)t^{deg}$ then p is the Alexander polynomial of infinitely many inequivalent prime knots.

The hard part of proving this theorem is finding the first knot with polynomial equal to p. From this knot it is easy to construct many others by inserting smaller knots (or tangles) into the diagram.

5.2 The Jones polynomial and Kauffman bracket

It is somewhat surprising (particularly with hindsight) that it was not until 1984 that the Jones polynomial, formally very little different from the Alexander polynomial, was discovered and sparked off the discovery of a host of similar polynomial invariants such as the Kauffman bracket, Homfly and Dubrovnik polynomials. There exist several excellent, easily accessible surveys of what has been done in this field over the last few years, see for example Lickorish (1988) or Kauffman (1990), so here we will just concentrate on aspects of knot polynomial theory which are of particular interest in combinatorics. This is because of its intimate relationship with classical combinatorial concepts such as the chromatic and flow polynomials.

Although it is not the way in which this polynomial was originally conceived, the polynomial V can be defined recursively and computed from the following rules:

(5.2.1) $t^{-1}V_K(\asymp) - tV_K(\asymp) = (t^{1/2} - t^{-1/2})V_K(\asymp)$

(5.2.2) $V_U = 1$

(5.2.3) V is an invariant of ambient isotopy.

Note that V, as defined, is a Laurent polynomial in the variable $t^{1/2}$; this is just an historical accident, and $t^{1/2}$ should merely be regarded as a symbol whose square is the symbol t.

We now concentrate on the Kauffman bracket polynomial. This is very close to the Jones polynomial and also to polynomials already familiar in combinatorics.

The *bracket polynomial* $[D]$ of a link L is obtained from any link diagram D of L by applying the equations

(5.2.4) $$[\asymp] = A[\asymp] + B[\supset \subset],$$

(5.2.5) $$[D U] = d[D],$$

(5.2.6) $$[U] = 1,$$

locally. It is not difficult to show that $[D]$ is well defined on link diagrams and is a polynomial in the 3 variables A, B and d, which are assumed to commute.

As it stands it is not an invariant of isotopy. However, suppose that we consider the second Reidemeister move. Applying the bracket rules we obtain the following formula.

$$[\asymp] = A[\asymp] + B[\asymp]$$

$$= AB[\asymp] + A^2[\supset \subset] + B^2[\supset \subset] + BA[\supset \circ \subset]$$

$$= (A^2 + B^2 + ABd)[\supset \subset] + AB[\asymp].$$

But in order for the bracket to be invariant under isotopy, it certainly must satisfy

$$[\asymp] = [\asymp]$$

and this forces the relations

$$AB = 1, \quad A^2 + B^2 + ABd = 0.$$

Thus if we specialise the bracket by insisting that

$$B = A^{-1}, \quad d = -(A^2 + A^{-2})$$

we have shown that the new one variable bracket of D, denoted by $\langle D \rangle$, is invariant under Reidemeister move II.

We now use this to show that the bracket is also invariant under Reidemeister move III. For notational convenience we leave out the bracket signs.

$$\text{[diagram]} \quad = \quad A \;\text{[diagram]}\; + \; A^{-1} \;\text{[diagram]}$$

$$= \quad A \;\text{[diagram]}\; + \; A^{-1} \;\text{[diagram]}$$

Using Reidemeister II equivalence,

$$\text{[diagram]} \quad = \quad A \;\text{[diagram]}\; + \; A^{-1} \;\text{[diagram]}$$

$$= \quad A \;\text{[diagram]}\; + \; A^{-1} \;\text{[diagram]}$$

again using Reidemeister II equivalence. This shows

$$\text{[diagram]} \quad = \quad \text{[diagram]}$$

and hence $\langle \;\; \rangle$ is invariant under Reidemeister II and III, in other words is an invariant of *regular isotopy*.   □

It is *not* an invariant under Reidemeister move I. However, provided that the link is oriented and then suitably normalised, Kauffman (1987a) showed that the resulting invariant is in fact the Jones polynomial of the link. More precisely, define the *writhe* $w(L)$ of an *oriented* link L to be the sum of the

+ve   -ve

Figure 5.4.

signs at crossings, calculated according to the convention shown in Figure 5.4, and then define a (Laurent) polynomial f on oriented link diagrams by

$$f_D(A) = (-A^3)^{-w(D)} \langle D \rangle$$

where $w(D)$ is the writhe of D and where $\langle \ \rangle$ for an oriented diagram is obtained by forgetting about the orientation.

(5.2.7) Proposition. The polynomial f is invariant under Reidemeister moves I, II and III.

Proof. The writhe is unchanged by moves II and III. Hence since the bracket is also invariant under these moves, so is f. Consider now move I. Let D be a link diagram with a portion as shown

$$D \quad \rlap{\quad\text{(loop diagram)}}$$

and let D' be obtained from D by removing the loop. Then

$$
\begin{aligned}
\langle D \rangle &= A \langle \ \rangle + A^{-1} \langle \ \rangle \ \bigcirc \ \rangle \\
&= (A + A^{-1}(-A^2 - A^{-2})) \langle D' \rangle \\
&= -A^{-3} \langle D' \rangle.
\end{aligned}
$$

However, no matter which way we orient the string in D the writhe of the crossing is $-ve$.

-ve   -ve

Hence $w(L) = w(L') - 1$. Thus $f(L) = f(L')$. The proof for the other loop is analogous. $\qquad \square$

But now it is immediate that f is nothing more than the Jones polynomial with a change of variable. In other words we can prove

(5.2.8). If K is an oriented link, and D an oriented diagram of K, then

$$V_K(t) = f_D(t^{-1/4}).$$

This leads immediately to:

(5.2.9) Theorem   The Jones polynomial $V_L(t)$ of an oriented link L is given by

$$V_L(t) = (-t)^{-3w(L)/4} \langle L \rangle_{A=t^{-1/4}}.$$

Proof.   Just verify that $f_D(t^{-1/4})$ satisfies the defining conditions of V_K.   □

Thus, for most purposes, the Kauffman bracket is as good an invariant as the Jones polynomial.

For example:

(5.2.10)   If L_1 and L_2 are isotopic then $\langle L_1 \rangle = A^\alpha \langle L_2 \rangle$, for some integer α.

In particular:

(5.2.11)   A necessary condition that L be isotopic to the unknot is that $\langle L \rangle$ is a power of A.

Another way of looking at the bracket polynomial, and one which is certainly easier in hand calculations is using signed graphs. Given a Tait colouring of a link diagram D form the associated signed graph as in §2.2. Now notice the actions on the graph corresponding to the operation of the bracket rules.

The rule

$$\langle \;\vartriangleright\!\!\vartriangleleft\; \rangle = A \langle \;)(\; \rangle + A^{-1} \langle \; \mathbb{D} \;\mathbb{C}\; \rangle,$$

on the link diagram corresponds to

$$\langle G \rangle = A \langle G_e'' \rangle + A^{-1} \langle G_e' \rangle$$

on the Tait graph of G. Here G_e', G_e'' denote the usual deletion and contraction of the edge e from G, where e is the edge corresponding to the crossing.

Also consider the evaluation of $\langle \;\; \rangle$ for a positive coloop and loop.

$$\langle \;\bowtie\!\!\!\bullet\; \rangle = A \langle \;\supset\; \rangle + A^{-1} \langle \;)\,\bigcirc\; \rangle$$

$$= A + A^{-1}(-A^2 - A^{-2}) \langle \;\supset\; \rangle$$

$$= -A^{-3}. \langle \;\supset\; \rangle$$

Similarly

$$\langle \;\bowtie\; \rangle = A(-A^2 - A^{-2}) + A^{-1} \langle \;\supset\; \rangle$$

$$= -A^3. \langle \;\supset\; \rangle$$

Thus the bracket rules are equivalent to

(5.2.12) $\qquad \langle G \rangle = A \langle G'_e \rangle + A^{-1} \langle G''_e \rangle$ if e is positive

(5.2.13) $\qquad \langle G \rangle = A \langle G''_e \rangle + A^{-1} \langle G'_e \rangle$ if e is negative,

where e is the edge joining the black faces in each case.

Combining these two rules with the rules $\langle C^+ \rangle = -A^{-3}$, $\langle C^- \rangle = -A^3$, $\langle L^+ \rangle = -A^3$, $\langle L^- \rangle = -A^{-3}$ where $C^\pm, L^\pm$ represent the positive (negative) signed isthmus (loop) respectively and we see that the bracket polynomial can be regarded as a "Tutte polynomial" on signed graphs.

This idea was made precise by Thistlethwaite (1987) and was one of the keys to his proof of Tait's conjecture as we see in §5.5.

In particular, we know from (2.3.1) that if L is an alternating link it has a representation as a link diagram in which each edge has a positive sign. Hence, comparison of the equations (5.2.12)-(5.2.13) with the defining relations of the Tutte polynomial given in §3.2 show that we have proved:

(5.2.14) Proposition. If L is an alternating link then its bracket polynomial is given by the relation

$$\langle L \rangle = A^{2|V(G)|-|E(G)|-2} T(G; -A^{-4}, -A^4)$$

where G is the unsigned graph of any positive link diagram representing L.

Example 1. The left hand trefoil, has positive graph G as shown

$$G = \quad \text{}$$

Since $T(G; x.y) = x + y + y^2$ the left hand trefoil has bracket given by

$$A^7 - A^3 - A^{-5}.$$

Thus for the left hand trefoil with orientation as shown in Figure 5.5, note that $w(D) = -3$, so

$$
\begin{aligned}
V(L) &= (-A)^9 (A^7 - A^3 - A^{-5}) \\
&= -A^{16} + A^{12} + A^4.
\end{aligned}
$$

Using $A = t^{-1/4}$, we get

$$V_L(t) = -t^{-4} + t^{-3} + t^{-1}.$$

$\qquad\qquad\qquad\qquad\qquad\qquad\qquad\qquad\qquad\qquad\qquad\qquad\qquad\quad\square$

Example 2. The right hand trefoil has associated graph G shown in Figure 5.6. It is easy to check that

$$T(G; x, y) = x^2 + x + y,$$

Figure 5.5.

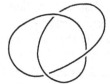

Figure 5.6.

Hence

$$\langle \text{Right trefoil} \rangle_A = A^{6-3-2}(+A^{-8} - A^{-4} - A^4)$$
$$= -A^5 - A^{-3} + A^{-7}.$$

In other words, its bracket polynomial equals that of its mirror image, the left trefoil, evaluated at $A = A^{-1}$.   □

This is a special case of the general result.

(5.2.15) The bracket polynomials of a link and its mirror image are related by

$$\langle L; A \rangle = \langle \bar{L}; A^{-1} \rangle.$$

It is useful to give an alternative *states model* representation of $\langle L \rangle$ in terms of concepts more familiar in combinatorics. This is contained in the following.

(5.2.16) Proposition.   Let D be any link diagram representing the link L and suppose that the associated signed graph $G = G(D)$ is connected. For any subset $A \subseteq E(G)$, let A^+ and A^- denote the positive (negative) signed part respectively. Then with r denoting the rank function in G we have

(5.2.17) $\langle D \rangle = A^{|E^-|-|E^+|}(-A^2 - A^{-2})^{r(G)} \times$
$$\times \sum_{X \subseteq E} A^{4(r(X)-|X^-|)}(-A^4 - 1)^{r(G)+|X|-2r(X)}$$
$$= A^{|E^-|-|E^+|+2r(G)} \times$$
$$\times \sum_{X \subseteq E} A^{-4(r(X)-|X^+|)}(-A^{-4} - 1)^{r(G)+|X|-2r(X)}.$$

Proof: See Schwärzler and Welsh (1993). □

We consider now some elementary evaluations of the bracket polynomial. Equation (5.2.17) gives

$$\langle D; A = 1 \rangle = \sum_{X \subseteq E(G)} (-2)^{r(G)+|X|-2r(X)},$$

which equals $T(G; -1, -1)$, where G is the unsigned version of G. Using the known evaluation of the Tutte polynomial at this point (3.7.7) we get:

(5.2.18) The evaluation of $\langle D \rangle$ when $A = 1$ is independent of signing and is given by

$$\langle D; 1 \rangle = (-1)^{r(G)} 2^{\dim(\mathcal{C} \cap \mathcal{C}^*)},$$

where $\mathcal{C}, \mathcal{C}^*$ are respectively the cycle space and cocycle space of G.

When G is a planar graph it is known that $T(G; -1, -1) = (-1)^{|E(G)|}$ $(-2)^{k(m(G))-1}$, where $m(G)$ is the medial graph of G. As a corollary we get the following property of the Jones polynomial:

(5.2.19) For any link L, $V_L(1) = (-2)^{c(L)-1}$, where $c(L)$ is the number of components of L.

Other basic properties of the Jones polynomial $V_L(t)$ are the following.

(5.2.20) $V_L(-1) = \Delta_L(-1)$ where Δ is the normalised Alexander polynomial satisfying $\Delta(1) = 1$ and symmetric in t and t^{-1}.

(5.2.21) If L is a knot then $V_L(e^{2\pi i/3}) = 1$.

(5.2.22) If K is a knot then $dV_K(t)/dt_{t=1} = 0$.

(5.2.23) If $\tilde{L}$ denotes the mirror image of L then

$$V_{\tilde{L}}(t) = V_L(1/t).$$

(5.2.24) The connected sum $L_1 \# L_2$ of L_1 and L_2 is given by

$$V_{L_1 \# L_2}(t) = V_{L_1}(t) V_{L_2}(t).$$

[Note: a certain care with orientations is needed here.]

We close by emphasising that there are many examples of knots having the same bracket (and Jones) polynomial but which are not isotopic. Indeed there are links with the same Jones polynomial and different Alexander polynomials, see for example Anstee, Przytycki and Rolfsen (1989). A stronger invariant is the 2-variable Homfly polynomial which we discuss in the next section. It contains both the Alexander and Jones polynomial as specialisations.

5.3   The Homfly polynomial

The discovery of the Jones polynomial led to the discovery of several other such invariants. In this section we describe briefly two of these, for more details we refer to Kauffman (1990) and Lickorish (1988). Our description is in terms of their defining "skein relations". These relations are the knot counterparts of the contract/delete formulae of matroids and graphs, and in each case represent local changes at a crossing of the knot or link which can be carried out in an arbitrary order to give the same polynomial. Moreover each of these changes will have been designed so as to preserve equivalence under the Reidemeister moves, so that links which are ambient isotopic have the same polynomial.

The *Homfly polynomial* $P(L; l^{\pm 1}, m^{\pm 1})$ is a 2-variable Laurent polynomial in $l^{\pm 1}, m^{\pm 1}$ and defined for *oriented* links by:

(i) P (unknot) $= 1$,

(ii)   If L_+, L_-, L_0 are defined by

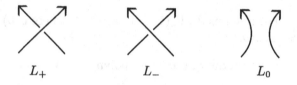

then

$$lP(L_+) + l^{-1}P(L_-) + mP(L_0) = 0.$$

(iii) P is well defined on ambient isotopy classes.

Example: As a very easy example consider the diagrams

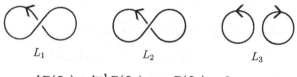

$$lP(L_1) + l^{-1}P(L_2) + mP(L_3) = 0.$$

Therefore since both L_1 and L_2 are ambient isotopic to the unknot,

$$P(L_3) = -m^{-1}l - m^{-1}l^{-1}.$$

$\square$

Other basic properties of P which are easily proved from the definition are the following.

(5.3.1) If L' is obtained from L by reversing the orientation of each component then $P(L') = P(L)$.

(5.3.2) If $\bar{L}$ is the mirror image of L then
$$P(\bar{L}; l, m) = P(L; l^{-1}, m).$$

The Homfly polynomial is a good but not infallible test of amphicheirality. There exist knots with crossing number 9 which are not achiral but have the same polynomial.

(5.3.3) If $L_1 \cup L_2$ denotes the disjoint union of L_1, L_2 then
$$P(L_1 \cup L_2) = -(l + l^{-1})m^{-1}P(L_1)P(L_2).$$

(5.3.4) The connected sum $L_1 \# L_2$ has Homfly polynomial given by
$$P(L_1 \# L_2) = P(L_1)P(L_2).$$

(5.3.5) The Alexander polynomial is given by
$$\Delta_L(t) = P(L; i, i(it^{1/2} - it^{-1/2}))$$
where $i^2 = -1$.

(5.3.6) The Jones polynomial is given by
$$V_L(t) = P(L; it^{-1}, i(t^{-1/2} - t^{1/2})).$$

   Note. Calculating the Homfly by "switching to the unknot" by a succession of switches is time consuming but it is essentially the only known way of calculating it.

Mutants and tangles
A *tangle* is a portion of a link diagram from which there emerge just 4 arcs pointing in the compass directions NW, NE, SW, SE, as typified in Figure 5.7. Tangles were first introduced by Conway (1969) and have led to a very beautiful algebraic theory based on their laws of composition. Mutation of a link is obtained by locating a tangle, and in the words of Anstee, Przytycki and Rolfsen (1989) "flipping it over like a pancake, or rotating it 180°". A precise definition is given in Lickorish and Millett (1987), but the idea can be seen from the example shown in Figure 5.8.
   Mutants can be formed in 3 ways from a given tangle/link diagram pair and correspond to the graph operation of "Whitney twist" by which pairs of 2 isomorphic graphs are related.
   A straightforward application of the recursive definition of the Homfly polynomial shows:

(5.3.7) If L' is obtained from L by mutation then $P(L') = P(L)$.

An immediate consequence is that:

(5.3.8) Mutants have the same Alexander and Jones polynomial.

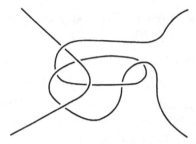

Figure 5.7.

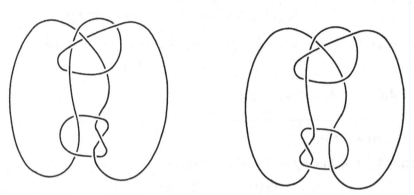

Figure 5.8. Famous mutants: the Conway and Kinoshita-Terasaka knots which both have trivial Alexander polynomial.

5.4 The Kauffman 2-variable polynomial

Kauffman's 2-variable polynomial $\Lambda(D; a, z) \in Z[a^{\pm 1}, z^{\pm 1}]$ is a Laurent polynomial of regular isotopy of a diagram D of an unoriented knot or link. In other words it is invariant under Reidemeister moves II and III. It is defined by the following formulae.

(5.4.1) For a simple closed curve O, $\Lambda(O) = 1$.

(5.4.2) $a^{-1}\Lambda(C_+) = a\Lambda(C_-) = \Lambda(C_0)$ where the C_i are as shown

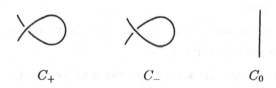

$$C_+ \qquad\qquad C_- \qquad\qquad C_0$$

(5.4.3) $\Lambda(D_+) + \Lambda(D_-) = z(\Lambda(D_0) + \Lambda(D_\infty))$ where the D_i are as shown.

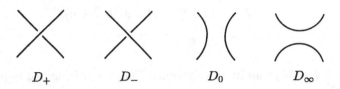

$$D_+ \qquad D_- \qquad D_0 \qquad D_\infty$$

To illustrate how Λ is computed we will obtain some of its elementary properties.

(5.4.4)
$$\Lambda(\bigcirc\ \bigcirc) = (a + a^{-1})z^{-1} - 1.$$

Proof.

$$\Lambda(\bigcirc\hspace{-0.3em}\infty) + \Lambda(\infty\hspace{-0.3em}\bigcirc) = z\Lambda(\bigcirc\ \bigcirc) + z\Lambda(\ \bigcirc\ )$$

Hence

$$z\Lambda(\bigcirc\ \bigcirc) = a^{-1} + a - z.$$

which proves (5.4.4). □

(5.4.5)
$$\Lambda(K \cup L) = (a^{-1} + a - z)z^{-1}\Lambda(K)\Lambda(L).$$

Proof. By induction on the number of crossings.

(5.4.6)
$$\Lambda(K_1 \# K_2) = \Lambda(K_1)\Lambda(K_2),$$

(5.4.7)
$$\Lambda(\bar{K}; a, z) = \Lambda(K; a^{-1}, z).$$

Example: We leave out brackets when there is no chance of confusion.

$$\Lambda\ \bigcirc\hspace{-0.6em}\bigcirc\ + \Lambda\ \bigcirc\hspace{-0.6em}\bigcirc$$

$$= z(\Lambda\ \bigcirc\hspace{-0.5em}\bigcirc\ + \Lambda\ \bigcirc\hspace{-0.5em}\bigcirc\ )$$

$$= z(a^{-1}\Lambda\ \bigcirc\ + a\Lambda\ \bigcirc\ )$$

$$= za^{-1} + za.$$

Now

$$\Lambda(L_1 \cup L_2) = ((a^{-1} + a)z^{-1} - 1)\Lambda(L_1)\Lambda(L_2)$$
$$\Lambda(L_1 \# L_2) = \Lambda(L_1)\Lambda(L_2).$$

Hence

$$\Lambda(\ \bigcirc\hspace{-0.6em}\bigcirc\ ) = za^{-1} + za - (a^{-1} + a)z^{-1} + 1.$$ □

Exercise. Show that Λ of the left handed trefoil is

$$-2a^{-1} - a + (1+a^2)z + (a^{-1}+a)z^2.$$

□

The 2-variable Kauffman polynomial $F(L; a, z)$ of oriented isotopy is then defined by

$$F(L; a, z) = a^{-w}\Lambda(D; a, z),$$

where w is the writhe of the oriented link L formed by the orientation of D.

Recall that the *writhe* w of an oriented link is the sum of the signs of the crossing points, according to the convention shown in Figure 5.5.

Note 1: There is no "easy" known proof of the fact that this is an oriented link invariant.

Note 2: F is sometimes called *semi-oriented* because, on changing the orientation of one component to get L^*,

$$F(L^*) = a^{4\lambda}F(L)$$

where λ is the linking number of the component with the rest of L.

The relationship between Λ or (almost equivalently) F and other polynomials we have already considered is summarised in

(5.4.8) $\langle D \rangle = \Lambda(D; -A^3, A+A^{-1})$.

(5.4.9) $V_L(t) = F(L; -t^{3/4}, t^{-1/4} + t^{1/4})$.

Elementary properties of F are the following:

(5.4.10) $F(L_1 \# L_2) = F(L_1)F(L_2)$ where $L_1 \# L_2$ is any connected sum of L_1 and L_2.

(5.4.11) $F(\bar{L}; a, z) = F(L; a^{-1}, z)$.

(5.4.12) If L_2 is a mutant of L_1, then $F(L_1) = F(L_2)$.

It should also be emphasised that there exist non-isotopic links with the same Homfly polynomial but different Kauffman polynomials and vice versa.

However, although F and Λ are quite strong invariants of links it is known that there are links which are not isotopic but which have the same Kauffman polynomial, though of course even more strongly than for the Jones polynomial the following question is open

(5.4.13) Question: Does there exist a knot K which is not equivalent to the unknot and for which $\Lambda(K) = 1$?

Figure 5.9.

Finally we mention an interesting one variable specialisation of Λ, this is the Q-polynomial invariant of unoriented links introduced by Brandt, Lickorish, Millett and Ho and which we will define by

(5.4.14) $$Q_L(x) = \Lambda(L; 1, x).$$

This was not the way it was originally conceived, indeed it could well be regarded as the 1-variable predecessor of Λ.

A very useful result of Kidwell (1987) is the following

(5.4.15) Theorem. If D is a connected link diagram for L then the degree of Q_L is not more than $n - b(D)$, where n is the number of crossings and $b(D)$, the bridge number, is the maximum number of consecutive overpasses which occur anywhere in D.

5.5   The Tait conjectures

We now use the representation of the bracket obtained in §5.2 to settle some problems which have remained open since the last century and have come to be known as the "Tait conjectures". For an account of their history we refer to the articles of de la Harpe, Kervaire and Weber (1986) and Thistlethwaite (1985).

(5.5.1) Theorem. If a link L has an alternating diagram D which is reduced and has n crossings then there is no diagram representing L which has fewer than n crossings.

A diagram is *reduced* if it has no crossing of the type shown in Figure 5.9.

In other words; given an alternating reduced diagram D of L we know that the crossing number of L is the number of crossings in D.

Proof. The key idea of the proof is to notice that since $(A^3)^{-w(D)}\langle D \rangle$ is an invariant of isotopy, we know that span $\langle D \rangle$ is also an invariant of isotopy. (The *span* of a Laurent polynomial is the difference between its maximum and minimum degrees.)

Now since D is an alternating link diagram, we may assume that the black face graph G defined by its Tait colouring, has only positive edges. Otherwise take the dual.

But using the representation

$$\langle D \rangle = A^\alpha T(G; -A^{-4}, A^4)$$

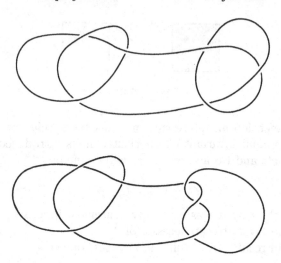

Figure 5.10. Alternating and nonalternating representations of the same knot.

and the representation of T as a rank polynomial it is easy to verify that

$$\text{span}\langle D\rangle = 4(r(E) + r^*(E))$$
$$= 4|E|,$$

provided that G has no loops or coloops. This is exactly the condition that the coefficients $t_{r,0}$ and t_{0,r^*} are non-zero. In other words, provided D is reduced

(5.5.2) $$\text{span}\langle D\rangle = 4|E(G)|$$

which in turn equals the number of crossings of D. But span $\langle D\rangle$ is an invariant of ambient isotopy. Hence (5.5.2) must hold for *any* reduced alternating representation of the link L. This proves Tait's First Conjecture. □

A stronger version of Theorem (5.5.1) is the following.

(5.5.3) Theorem. Suppose that L is an alternating prime link. Then any nonalternating diagram representing L has crossing number strictly greater than the span of V_L.

To see that the property of being prime is needed above, consider the two representations of the granny knot as the direct sum of the right and left trefoil shown in Figure 5.10

This construction can be extended using any pair of alternating knots K_1, K_2. All one does is to obtain two alternating representations of K_2, one having a positive Tait graph G_2 the other having a negative Tait graph G_2^*, the plane dual of G_2. Now form $K_1 \# K_2$ in two distinct ways, corresponding to this direct sum. The one from G_1, G_2 will be an alternating representation the other from G_1, G_2^* will be nonalternating.

Tait's Evenness Conjecture

Tait and his coworkers of the nineteenth century seemed to discount the existence of amphicheiral links of odd crossing number. It is still not known whether any exist.

What does follow from the above is the following statement.

(5.5.4) Theorem. Every alternating amphicheiral link has even crossing number.

Proof. If L is isotopic to its mirror image, then since $\langle L; A \rangle = \langle \bar{L}; A^{-1} \rangle$, span $\langle L \rangle$ is an even integer. But for alternating links, we have just shown that span $\langle L \rangle$ is the crossing number. $\qquad \square$

Another obvious consequence of Theorem 5.5.1 is:

(5.5.5) If $V_K(t) = 1$ and K is alternating, then K is the unknot.

Hence any link with $V_K(t) = 1$ and K not the unknot must be non-alternating, moreover since all nonalternating links on fewer than 14 crossings are known to have nontrivial Jones polynomial, any example which shows that $V_K(t) = 1$ does not characterise the unknot must have 14 or more crossings.

Tait's Flyping Conjecture

In contrast to the two previous conjectures whose proofs (with hindsight) are relatively easy, Tait's Flyping Conjecture which we now address is of a different order of magnitude.

A *flype* is a move on a tangle (with 2 input and 2 output strings) obtained by rotating the tangle by 180°.

Figure 5.11. A flype.

The *flyping conjecture* of Tait, made at the end of the last century, is that any two reduced alternating projections of the same link can be obtained from each other by a sequence of flypes.

Now Thistlethwaite (1988) observes that if L, K are two alternating links related by a flype then their positive graphs $G(L)$ and $G(K)$ satisfy

$$(5.5.6) \qquad T(G(L); x, y) = T(G(K); x, y).$$

In other words the Tutte polynomial is invariant under flypes. This is an immediate consequence of the following fact.

(5.5.7) If K and L are related by a flype then $G(K)$ and $G(L)$ are 2-isomorphic.

In 1990 L. Schrijver constructed a proof that Tait's flyping conjecture was true for links which had well connected link diagrams. A diagram is *well connected* if it has no 2-edge cut set and the only 4-edge cut sets are those consisting of the edges incident with a common vertex. This was a major advance and it was not too surprising when the full flyping conjecture was proved correct in 1991 when Menasco and Thistlethwaite announced their proof. Neither of these proofs is easy, they involve intricate geometric reasoning and not even a sketch is given here.

5.6   Thistlethwaite's nontriviality criterion

Thistlethwaite (1988) has found a striking combinatorial relation linking the 2-variable Kauffman polynomial Λ of a link L with the chromatic and flow polynomials of certain minors of the checkerboard graph $G = G(D)$, where D is any diagram of the link. This also leads to a powerful test of nontriviality.

In order to be able to present the result we first need some preliminaries.

(5.6.1) If D has n crossings and

$$\Lambda(D; a, z) = \sum u_{rs} a^r z^s$$

then $u_{rs} \neq 0 \Rightarrow n + r + s$ is even.

Proof. Induction on the ordered pair (n, k) of D, where n is the number of crossings and k is the number of switches needed to change D to the unlink. These pairings are ordered lexicographically and the inductive step is achieved by the switching relation for Λ. $\square$

(5.6.2) If $u_{rs} \neq 0$ then $|r| + s \leq n$ where n is the number of crossings of D.

Proof: See Thistlethwaite (1988).

Using just (5.6.2) it is convenient to illustrate the scope of Λ by a "triangular" array.

Now define the two *outer polynomials* of D by

$$\phi_D^+(t) = u_{0,n} + u_{1,n-1}t + u_{2,n-2}^2 t^2 + \ldots + u_{n,0}t^n$$

$$\phi_D^-(t) = u_{0,n} + u_{-1,n-1}t + u_{-2,n-2}t^2 + \ldots + u_{-n,0}t^n.$$

If we now let G_+, G_- denote the subgraphs of G obtained by restricting to the sets of positive (respectively negative) edges of G, then we can state Thistlethwaite's theorem.

(5.6.3) Theorem. For any link diagram D, the outer polynomials are given by
$$\phi_D^+(t) = T(G_+; 0, t)T(G_-^*; 0, t)$$
$$\phi_D^-(t) = T(G_+^*; 0, t)T(G_-; 0, t)$$
where G is the signed graph $G(D)$.

Alternatively, in the terminology of graph minors
$$\phi_{D+}(t) = T(G \mid E^+; 0, t)T(G.E^-; t, 0)$$

$$\phi_{D-}(t) = T(G \mid E^-; 0, t)T(G.E^+; t, 0)$$
where $G \mid A$, $G.A$ denote respectively the deletion and contraction of the edges of $E(G) \backslash A$.

Using (3.7.8) and (3.7.9) we can restate Theorem 5.6.3 in the form

(5.6.4) Corollary. The outer polynomials of the link diagram D satisfy
$$\phi_{D+}(t) = k^+ F(G \mid E^+; \lambda)P(G.E^-; \lambda)$$

$$\phi_D^-(t) = k^- F(G \mid E^-; \lambda)P(G.E^+; \lambda)$$
where F and P are the flow (chromatic) polynomials, $\lambda = 1 - t$, and $k^{\pm}$ are easily determined constants.

An immediate consequence is:

(5.6.5) Corollary. Let D be a link diagram with at least one non-nugatory crossing, then if either

(5.6.6) $$P(G^+; t)F(G^-; t) \neq 0$$

or
(5.6.7) $$F(G^+; t)P(G^-; t) \neq 0,$$

then D represents a nontrivial link.

Note. This is a remarkably effective test inasmuch as all ≤ 11-crossing knot diagrams satisfy it. However there does exist a 12-crossing (nontrivial) knot which has no 12-crossing diagram satisfying either of these criterion.

It is tempting to conjecture from what has gone before that $P(G^+; t)$ $F(G^-; t)$ might be a link invariant. This is not the case.

It should also be noted that (5.6.5) leads directly to the following easily checked sufficient condition for nontriviality. A graph has non-zero chromatic polynomial unless it has a loop, similarly it has non-zero flow polynomial provided it has no isthmus. Hence the conditions (5.6.6–7) reduce to the following

(5.6.8) A link diagram represents a nontrivial link if its graph G is such that either

(i) $G \mid E^-$ has no isthmus and $G.E^+$ has no loop,

or

(ii) $G \mid E^+$ has no isthmus, and $G.E^-$ has no loop.

These turn out to correspond to the notions of *adequacy* and *semiadequacy* introduced by Lickorish & Thistlethwaite (1988), see also Schwärzler and Welsh (1992).

5.7   Link invariants and statistical mechanics

There is a clear similarity between the state representations of the various knot polynomials and the partition functions of statistical mechanics discussed earlier. This has been heavily exploited over the last few years, see for example Jones (1989), Turaev (1988) and Witten (1989). There is now considerable literature in this area and all I aim to do here is to give an introduction to the basic ideas. We will consider what I think is the simplest example as presented in Jones (1989).

(5.7.1) Definition: A *spin model* $S = \{\Theta, \omega_\pm\}$ is a set Θ of n "spins" and functions $\omega_\pm(a, b)$, $a, b \in \Theta$ such that for all $a, b, c \in \Theta$,

(5.7.2) $$\omega_\pm(a, b) = \omega_\pm(b, a),$$

(5.7.3) $$\omega_+(a, b)\omega_-(a, b) = 1,$$

(5.7.4) $$\sum_{x \in \Theta} \omega_-(a, x)\omega_+(x, c) = n\delta(a, c)$$

where δ is the Kronecker delta,

(5.7.5) $$\sum_{x \in \Theta} \omega_+(a, x)\omega_+(b, x)\omega_-(c, x) = \sqrt{n}\,\omega_+(a, b)\omega_-(b, c)\omega_-(c, a).$$

Given a signed graph $G = \langle V, E \rangle$ define a *state* σ to be any function $\sigma : V \mapsto \Theta$ and the *partition function* Z_G^S is defined by

$$Z_G^S = \left(\frac{1}{\sqrt{n}}\right)^{|V|-1} \sum_\sigma \prod_{e \in E} \omega(e)$$

where the sum is over all states σ and if $e = (i, j)$, $\omega(e) = \omega_\pm(\sigma(i), \sigma(j))$ depending on the sign of e.

In particular if G is the plane signed Tait graph of the link diagram D we write Z_D^S for Z_G.

(5.7.6) Theorem. For any spin model S, and connected link diagrams, D_1 and D_2, if the corresponding links L_1 and L_2 are regularly isotopic, then $Z_{D_1}^S = Z_{D_2}^S$.

The proof is straightforward and depends upon noticing that the condition (5.7.5) is designed to capture star-triangle equivalence, shown below which by Theorem 2.2.3 is equivalent to Reidemeister III.

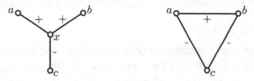

The normalisation factor $(1/\sqrt{n})^{|V|}$ is to account for the change in the number of vertices of the Tait graph under Reidemeister III. The condition (5.7.4) forces Reidemeister II to hold.

The idea now is to find solutions in ω of the equations (5.7.2)- (5.7.5). These will then give invariants of regular isotopy and as has been emphasised by Kauffman in his series of papers, this is the crucial component of ambient isotopy.

Jones points out, for example, that:

(5.7.7) When $n = 2$ or 3 the only invariants obtained are the evaluations $V_L(i)$ and $V_L(e^{i\pi/3})$ of the Jones polynomial.

(5.7.8) When $n = Q$ and ω_+ is given by

$$\omega_+(a,b) = \begin{cases} 1 & \text{if} \quad a = b \\ -t^{-1} & \text{if} \quad a \neq b, \end{cases}$$

we get a solution provided t satisfies the equation

$$2 + t + t^{-1} = Q.$$

In this case the resulting invariant is essentially the Kauffman bracket polynomial.

(5.7.9) Metaplectic invariants: Suppose that Θ is a group H and we look for solutions satisfying the additional condition that $\omega_+(a,b)$ is a function of ab^{-1}. Then a solution is

$$f(p) = K e^{\pi i p^2/n}$$

where $K^{-2} = \sum_{a \in Z/nZ} e^{2\pi i a^2/n}$.

There is a correspondence between spin models and what are known as IRF Models in which the energy of a state is the sum of the energies of each face, and the energy of a face is determined by the states of its vertices.

In turn, IRF models are closely related to what are known as vertex models in which spins (or colours) are assigned to the edges of an oriented link diagram and the resulting conditions of regular isotopy of *oriented* Reidemeister moves which correspond to (5.7.2) - (5.7.5) above are essentially equivalent to the Yang-Baxter relation. The equations become much more complicated, see for example Witten (1989). However the basic idea remains the same, namely that integrability of a lattice model, invariance under Reidemeister moves and satisfying Yang-Baxter are very closely related.

5.8   Additional notes

The treatment of the Alexander polynomial is based on Kauffman (1987b) and Thistlethwaite (1985). There are several different representations of the Alexander polynomial as a determinant. These use the Wirtinger presentation or the Bureau representation of the associated closed braid. For details see Burde and Zieschang (1985) or Crowell and Fox (1977). Thus the Alexander polynomial can be computed in polynomial time, in contrast to each of the other polynomials considered here whose computation is inherently exponential.

The discovery of the Homfly polynomial was made independently by different researchers shortly after the discovery of the Jones polynomial, see the announcement by Freyd, Yetter, Hoste, Lickorish, Millett and Ocneanu (1985). The parametrisation which we use follows that of Lickorish (1988), the transformation $l = it^{-1}$, $m = ix$ where $i^2 = -1$ gives the variables used by Jones (1987). The term "pretzel knot" goes back to Reidemeister (1935).

Note that we have not given proofs that the Homfly and Kauffman 2-variable polynomial are invariant under isotopy. These can be found in the original papers - in particular Lickorish and Millett (1987) and Kauffman (1990).

The first Tait conjecture Theorem 5.5.1 was proved independently and by different methods by Murasugi (1987) and Thistlethwaite (1987). The latter's proof is given here. There exist several other link polynomials which for reasons of space we have not even mentioned here. The Akutsu-Wadati polynomials are of particular significance since they are an example of new knot invariants derived from exactly solvable models in statistical mechanics. For details of these and a very clear treatment of the relationship between knot polynomials and statistical mechanics see the very recent survey of Wu (1992). Atiyah (1990) is a beautiful account of more advanced topics on this frontier between geometry and physics. Jones (1991) is a concise elegant account of the relationship between knot polynomials, Hecke algebras and the theory of subfactors.

6 Complexity Questions

6.1 Computations in knot theory

One of the earliest reported applications of computers to knot theory consisted of calculating Alexander polynomials of knots with up to 10 crossings by Anger (1959), see Trotter (1969). Nowadays it is possible to obtain Alexander polynomials of very large knot diagrams. However, because of the $\#P$-hardness results to be described later there is little hope of being able to compute the Jones polynomial of similar size diagrams unless they have some special features.

For example Jones (1987) shows that given a closed braid representation of L, say $L = \hat{\alpha}$, where $\alpha \in B_m$, then P_L can be computed quite rapidly by the following method.

Write the braid α as a word in the σ_j and expand the product in terms of the basis of the associated Hecke algebra $H(q, m)$. From this representation one can calculate the trace $tr(\alpha)$ in time which is a linear function of the length of the word representing α. Then use the fact that the Homfly polynomial is given (up to an easily obtained factor) by $tr(\alpha)$.

The snag in this method is that the Hecke algebra $H(q, m)$ has dimension $m!$ Thus this approach will be practical only for links presented as braids with not too many strings. For example, Morton and Short (1990) have implemented a version of the above algorithm on an IBM 3083 which will deal with 8-string braids of up to 150 crossings. This restriction to braids of "narrow width" is reminiscent of similar restrictions for computing the Tutte polynomial in Oxley and Welsh (1992) and of the general theory of Robertson and Seymour which can be loosely described as "almost everything which is computable, is computable in polynomial time for structures of bounded tree width".

We will explain below why, unless something quite unexpected is the case, such as, that $\#P = FP$, computing the Jones, Homfly and Kauffman polynomials is an inherently exponential process.

Finally, it is worth highlighting again the problem of putting an upper bound on the number of Reidemeister moves needed to reduce an n-crossing diagram of the unknot to one with zero crossings. The fundamental question is the following:

(6.1.1) Problem. Find a function f such that if D is an n-crossing link diagram of the unknot then D can be transformed to a diagram with zero crossings in fewer than $f(n)$ Reidemeister moves?

As for the general question of knot equivalence, Paterson and Razborov (1991) have recently proved the following result which may have some implications. Consider the computational problem:

NON-MINIMAL BRAIDS
Instance: A braid group B and a word w, in the standard generators of B.
Question: Is there a shorter word w' which is equivalent to w in B?

(6.1.2) Theorem: *NON-MINIMAL BRAIDS* is NP-complete.

Note: Unlike the majority of NP-completeness results, perhaps the more surprising aspect of this result is that *NON-MINIMAL BRAIDS* belongs to NP. Artin's original algorithm for the word problem in the braid group involves generating a canonical form which is exponential in the length of the original word. However, due to recent (as yet unpublished) work of Thurston (1988) there is a polynomial time algorithm for the word problem in the braid group and this can be used to show that *NON-MINIMAL BRAIDS* belongs to NP.

6.2   The complexity of the Tutte plane

We have seen that along different curves of the x, y plane, the Tutte-Whitney polynomial evaluates such diverse quantities as percolation probabilities, the weight enumerator of a linear code, the partition function of the Ising and Potts models of statistical physics, the chromatic and flow polynomials of a graph, and the Jones and Kauffman bracket polynomials of an alternating knot. Since it is also the case that for particular curves and at particular points the computational complexity of the evaluation can vary from being polynomial time computable to being $\#P$-hard, a more detailed analysis of the complexity of evaluation is needed in order to give a better understanding of what is and is not computationally feasible for these sort of problems. The section is based on the paper of Jaeger, Vertigan and Welsh (1990) which will henceforth be referred to as [JVW].

First consider the problem:

$\pi_1[\mathcal{C}]$: *TUTTE POLYNOMIAL OF CLASS $\mathcal{C}$*
Instance: Graph G belonging to the class $\mathcal{C}$.
Output: The coefficients of the Tutte polynomial of G.

We note first that for all but the most restricted classes this problem will be $\#P$-hard. This follows from the following observations.

(6.2.1) Determining the Tutte polynomial of a planar graph is $\#P$-hard.

Proof. Determining the chromatic polynomial of a planar graph is $\#P$-hard and this problem is the evaluation of the Tutte polynomial along the line $y = 0$.   □

It follows that:

(6.2.2) If C is any class of graphs which contains all planar graphs then $\pi_1[C]$ is $\#P$-hard.

However it does not follow that it may not be easy to determine the value of $T(G;x,y)$ at particular points or along particular curves of the x,y plane. For example, the evaluation of the Tutte polynomial at $(1,1)$ gives the number of spanning trees of the underlying graph and hence the Kirchhoff determinantal formula shows:

(6.2.3) Evaluating $T(G;1,1)$ for general graphs is in P.

Due to our interest in knots and the evaluation of the Jones polynomial, we are particularly interested in the evaluation of the Tutte polynomial of planar graphs along the hyperbola $xy = 1$. Accordingly we are led to the formulation of two further problems.

$\pi_2[C:L]$ *TUTTE POLYNOMIAL OF CLASS C ALONG CURVE L*

Instance: Graph G belonging to the class C.

Output: The Tutte polynomial along the curve L as a polynomial with rational coefficients.

$\pi_3[C:a,b]$ *TUTTE POLYNOMIAL OF CLASS C AT* (a,b)

Instance: Graph G belonging to the class C.

Output: Evaluation of $T(G;a,b)$.

Note: There are some technical difficulties here, inasmuch as we have to place some restriction on the sort of numbers on which we do our arithmetic operations, and also the possible length of inputs. Thus we restrict our arithmetic to be within a field F which is a finite dimensional algebraic extension of the rationals. We also demand (for reasons which become apparent) that F contains the complex numbers i and $e^{2\pi i/3}$. Similarly we demand that any curve L under discussion will be a rational algebraic curve over such a field F and that L is given in standard parametric form. For more details see [JVW] and for more on this general question see Grötschel, Lovász and Schrijver (1988).

Now it is obvious that for any class C, if evaluating T at (a,b) is hard and $(a,b) \in L$ then evaluating T along L is hard. Similarly if evaluating T along L is hard then determining T is hard. In other words:

(6.2.4) For any class C, curve L and point (a,b) with $(a,b) \in L$,

$$\pi_3[C;a,b] \propto \pi_2[C;L] \propto \pi_1[C].$$

Two of the main results of [JVW] are that except when (a, b) is one of a few very special points and L one of a special class of hyperbolae, then the reverse implications hold in (6.2.4).

Before we can state the two main theorems from [JVW] we need one more definition. Call a class of graphs *closed* if it is closed under the operations of taking minors and series and parallel extensions. That is C shall remain closed under the four operations of deletion and contraction of an edge together with the insertion of an edge in series or in parallel with an existing edge.

The first result of [JVW] relates evaluations in general with evaluation along a curve.

(6.2.5) Theorem. If C is any closed class then the problem $\pi_1[C]$ of determining the Tutte polynomial of members of C is polynomial time reducible to the problem $\pi_2[C; L]$ for any curve L, except when L is one of the hyperbolae defined by

$$H_\alpha \equiv (x - 1)(y - 1) = \alpha \qquad \alpha \neq 0,$$

or the degenerate hyperbolae

$$H_0^x \equiv \{(x, y) : x = 1\}$$
$$H_0^y \equiv \{(x, y) : y = 1\}.$$

We call the hyperbolae H_α *special hyperbolae* and an immediate corollary of the theorem is:

(6.2.6) Corollary. If L is a curve in the x, y plane which is not one of the special hyperbolae, and $\pi_1(C)$ is #P-hard then $\pi_1(C; L)$ is #P-hard.

Proof (sketch). Given a curve L and a class C we need to show that the existence of a polynomial time algorithm for determining $T(G; L)$ for $G \in C$ would lead to a polynomial time algorithm for determining the coefficients of $T(G; x, y)$. To do this we use a "tensor product" of matroids introduced by Brylawski (1980) to transform $M \in C$ to its tensor product $M \otimes N$, where N is also chosen so as to make $M \otimes N \in C$. In the case considered here C will be a closed class of graphic matroids but the argument is general. The tensor product $M \oplus N$ corresponds to "glueing" a copy of N to each element of M. The relation

$$T(M \otimes N; x, y) = \alpha T(M; X, Y)$$

with α an easily determined quantity and

$$X = ((x - 1)f + g)/f$$

(6.2.7)

$$Y = (f + (y - 1)g)/g$$

give the value of the Tutte polynomial of M at a succession of points in the plane as N varies.

The functions f, g depend on x, y, N and a particular point d of N and are given by the solutions to

$$(x - 1)f + g = T(N \backslash d; x, y)$$

$$f + (y - 1)g = T(N/d; x, y).$$

Thus by suitably varying N we can obtain evaluations of $T(M; X, Y)$ at a succession of points in the plane and then using Lagrange interpolation we can recover T everywhere. This method will work for all curves L except when L is one of the special hyperbolae; the reason for this being that inspection of (6.2.7) shows that it transforms the special hyperbola H_α into itself. □

The second main theorem of [JVW] relates the complexity of determining the Tutte polynomial along a curve with determining its value at a particular point on the curve.

(6.2.8) Theorem. The problem $\pi_3[\mathcal{C}; a, b]$ of evaluating $T(G; a, b)$ for members G of $\mathcal{C}$ (a closed class) is polynomial time reducible to evaluating T along the special hyperbola through (a, b) except when (a, b) is one of the special points $(1,1),(0,0),(-1,-1),(0,-1),(-1,0),(i, -i), (-i, i)$ and $(j, j^2), (j^2, j)$ where $i^2 = -1$ and $j = e^{2\pi i/3}$.

In other words unless a point is one of these 9 special points evaluating the Tutte polynomial at that point is no easier than evaluating it along the special hyperbola through that point.

Proof (outline). The underlying idea is similar to that used in proving Theorem (6.2.5). The transformations (6.2.7) are used to obtain the value $T(G; x, y)$ at a series of points along the hyperbola H_c, where $c = (a - 1)(b - 1)$. Thus provided the transformation moves (a, b) to enough points of the curve we will be able to recover the evaluation along the whole curve by Lagrange interpolation. The reason why the argument does not work for the 9 special points listed in the statement of the theorem is that the transformation switches them amongst themselves for all choices of N. □

The full details of this proof are rather technical and we refer to [JVW].

As an illustration of the applicability of these results we state without proof the following theorem from [JVW].

(6.2.9) Theorem. The problem of evaluating the Tutte polynomial of a graph at a point (a, b) is #P-hard except when (a, b) is on the special hyperbola

$$H_1 \equiv (x - 1)(y - 1) = 1$$

or when (a,b) is one of the 9 special points $(1,1),(-1,-1),(0,-1)$, $(-1,0),(i,-i)$, $(-i,i),(j,j^2)$ and (j^2,j), where $j = e^{2\pi i/3}$. In each of these exceptional cases the evaluation can be done in polynomial time.

The planar case

As far as planar graphs are concerned, there is a significant difference. The technique developed using the Pfaffian to solve the Ising and dimer problems for the plane square lattice by Fisher (1966) and Kasteleyn (1961) can be extended to give a polynomial time algorithm for the evaluation of the Tutte polynomial of any planar graph along the special hyperbola

$$H_2 \equiv (x-1)(y-1) = 2.$$

Thus this hyperbola is also "easy" for planar graphs. However it is easy to see that H_3 cannot be easy for planar graphs since it contains the point $(-2,0)$ which counts the number of 3-colourings and since deciding whether a planar graph is 3-colourable is NP-hard, we know that this must be at least NP-hard. However it does not seem easy to show that H_4 is hard for planar graphs. The decision problem is after all trivial by the four colour theorem. The fact that it is $\#P$-hard is just part of the following extension of Theorem (6.2.9) due to Vertigan (1991a).

(6.2.10) Theorem. The evaluation of the Tutte polynomial of a planar graph at a point (a,b) is $\#P$-hard except when

$$(a,b) \in H_1 \cup H_2 \cup \{(1,1),(-1,-1),(j,j^2),(j^2,j)\}$$

when it is computable in polynomial time.

More recently this has been further strengthened in the following result of Vertigan and Welsh (1992)

(6.2.11) Theorem: The conclusion of Theorem (6.2.10) holds for the class of bipartite planar graphs.

This suggests the natural question of whether by further restricting the class of objects more points in the "Tutte plane" can have polynomial time evaluations. From the following result of Oxley and Welsh (1992) there is not much room for manoeuvre since a special case of the main result of that paper is:

(6.2.12) Theorem. The Tutte polynomial of a series parallel network can be evaluated in polynomial time at any algebraic point (a,b) of the plane.

6.3   The complexity of knot polynomials

As we mentioned earlier the algorithmic question of deciding whether two knots are ambient isotopic seems to be difficult. Reidemeister's theorem gives no bound on the number of (Reidemeister) moves needed to decide whether or not a given knot diagram on n crossings represents the unknot. The known algorithms, see Haken (1961) and Hemion (1979), are not based on Reidemeister moves but on more complicated topological operations. As a result good knot invariants are extremely valuable as a means of separating non-equivalent knots.

The first polynomial invariant, the Alexander polynomial has long been of fundamental importance. It is easy to obtain, being just the expansion of a determinant. The later polynomials, such as the Homfly and 2-variable Kauffman bracket polynomial whose discovery followed from that of the Jones, are by their bivariate nature more time consuming than the Jones to calculate and it was shown by Jaeger (1988b) and Thistlethwaite (1987) that determining the Homfly polynomial of a knot was NP-hard. Both of these polynomials contain the Jones polynomial as a 1-variable specialisation, as do most of the other more recently discovered knot polynomials. Accordingly it was an open question whether or not the Jones, a single variable polynomial, was like the Alexander and computable in polynomial time or had the same characteristic as its bivariate extensions and was computationally hard. This is answered in the following theorem from Jaeger, Vertigan and Welsh (1992).

(6.3.1) Theorem. Determining the Jones polynomial of an alternating link is $\#P$-hard.

Proof. From §5.2 we know that determining the Jones polynomial of a link given an alternating link diagram $D(L)$ representing L is Turing equivalent to determining the Tutte polynomial of its underlying graph $G(L)$ along the hyperbola $xy = 1$. Since $xy = 1$ is not a special hyperbola as defined in Theorem (6.2.5), and since the set of underlying graphs associated with alternating link diagrams is the set of all planar graphs, it shows that determining the Jones polynomial is $\#P$-hard. $\qquad\square$

It is now natural to consider the evaluations of the Jones polynomial at particular values of the argument. In order to do this we recall that we are operating in the class of planar graphs, and since the hyperbola $xy = 1$ intersects the set of "easy points" listed in Vertigan's Theorem (6.2.8) in just eight points, we have

(6.3.2) Theorem. Evaluating the Jones polynomial $V_K(t)$ of an alternating link at any point t is $\#P$-hard except when t takes one of the eight special values $1, -1, -j, -j^2, i, -i, j, j^2$ when it can be evaluated in polynomial time.

Other knot polynomials

Both the Homfly polynomial P and the Kauffman polynomial F contain the Jones polynomial as specialisations, namely

$$(6.3.3) \qquad V_K(t) = P_K(L; it^{-1}, i(t^{-\frac{1}{2}} - t^{\frac{1}{2}})),$$

and

(6.3.4) $$V_K(t) = F_K(L; t^{-\frac{3}{4}}, -(t^{-\frac{1}{4}} + t^{\frac{1}{4}})).$$

Hence, they must be at least as hard to evaluate as the Jones polynomial. This means that there is unlikely to be a positive answer to the question raised by Lickorish and Millett (1988), namely can either the P or F polynomial be determined in subexponential time? From Theorem 6.3.1, we know that the answer to this is negative unless any problem in $\#P$ can be done in subexponential time, and this is most unlikely.

Vertigan (1991b) has carried out a detailed analysis of the points at which the Homfly and Kauffman polynomials have easy evaluations and obtained results analogous to those obtained in Theorem 6.3.2 for the Jones polynomial. In each case there is a finite collection of special curves and points for which the evaluation is easy. These include the curve corresponding to the Alexander polynomial and the easy points already found for the Jones. Almost all have known interpretations in the knot theory literature, though one slightly surprising feature is that in the case of the Kauffman polynomial there is a curve which is easy for knots but $\#P$-hard for links.

Vertigan's methods involve extremely intricate constructions and it is not possible to do more than present the results here.

(6.3.5) Theorem. Evaluating the Homfly polynomial $P(L; l, m)$ is $\#P$-hard at all points $(l.m)$ unless $l = \pm i$ or $m = \pm(l + l^{-1})$ or

$$(l, m) \in \{(\pm 1, \pm 1), (\pm 1, \pm\sqrt{2}), (\pm e^{\pm i\pi/6}, \pm 1)\}$$

where all the $\pm$ signs are independent. In each of these cases P is polynomial time computable.

(6.3.6) Theorem. Evaluating the Kauffman polynomial $F(L; w, z)$ is $\#P$-hard unless $w = \pm i$ or

(i) $(w, z) = (-q^{\pm 3}, q + q^{-1})$ where $q^{16} = 1$ or $q^{24} = 1$ but $q \neq \pm i$;   or

(ii) $(w, z) = (q^{\pm 3}, q + q^{-1})$ where $q^8 = 1$ or $q^{12} = 1$ but $q \neq \pm i$;   or

(iii) $(w, z) = (-q^{\pm 1}, q + q^{-1})$ where $q^{16} = 1$ but $q \neq \pm i$;   or

(iv) $(w, z) = \pm(1, q + q^{-1})$ where $q^5 = 1$.

In all these exceptional cases the polynomial is polynomial time computable.

What is particularly interesting about these results is that, like the earlier results in Theorem (6.3.2) about the Jones polynomial, all the points at which the relevant polynomial is polynomial time computable, are points at which knot theorists already had known interpretations.

Subexponential knot invariants

We conclude this section with a brief discussion of what is, in theory, practical when discussing knot invariants.

First note, that provided one accepts that $\#P$ is unlikely to equal FP, then it could be argued that the Alexander polynomial is better value as a knot invariant than any of the link polynomials which contain the Jones polynomial as a special case.

However, in a recent paper Przytycka and Przytycki (1992) have come up with the idea of using not the full Jones (or Homfly) polynomial as an invariant, but partial invariants which are provably computable in subexponential time.

For example, they show that the following is true.

(6.3.7) Theorem. If $V_L(t)$ is the Jones polynomial of L derived from a link diagram having m crossings then

$$V_L(t) \bmod ((t-1)^s)$$

can be computed in time $O(m^s)$.

Similarly, they show, that the following holds.

(6.3.8) If $P(L, a, z)$ and $F(L; w, z)$ denote the Homfly (respectively Kauffman) polynomial of L derived from a diagram of m crossings then

$$P(L; a, z) \bmod (z^s)$$

and

$$F(L; w, z) \bmod (z^s)$$

can be computed in $O(m^s)$ time.

We close this section by considering a question which if answered negatively would have profound consequences for the unknotting problem.

Recall first that the exact status in the computational complexity time hierarchy of the following problem is far from clear.

THE UNKNOTTING PROBLEM

Instance: A link diagram D on n crossings.

Question: Is D the link diagram of the unknot?

Probably the most important algorithmic question about the Jones polynomial is the following.

(6.3.9) Problem: Is there a nontrivial knot K for which $V_K(t) = 1$, other than the unknot?

If there were no such knot, we would have an algorithm for the unknotting problem, namely just calculate the Jones polynomial. By the standards of the existing algorithms this would be a major advance.

Note also that until this question is settled, none of the other knot polynomials which specialise to the Jones can characterise the unknot.

Using the relation between the Jones polynomial of an alternating link L and the Tutte polynomial of any associated graph $G(L)$ it is easy to prove:

(6.3.10) If L is an alternating link and $V_L(t) = 1$ then L is the unknot.

Thus the unknotting problem for alternating links is clearly contained in exponential time.

However there is still a problem; suppose L is presented as a nonalternating link diagram. If we calculate $V_L(t)$ and find $V_L(t) \neq 1$ then L is not the unknot; however if $V_L(t) = 1$ we are faced with two possibilities, *either* L is the unknot or L is a nonalternating knot with the same Jones polynomial as the unknot.

This prompts the following algorithmic question

(6.3.11) Problem: What is the status in the complexity hierarchy of the following question?

ALTERNATING KNOT

Instance: A link diagram D.

Question: Is D the link diagram of an alternating knot?

Clearly from what has gone before, we have

(6.3.12) $UNKNOTTING \propto ALTERNATING$,

but it seems difficult to say much more.

6.4   The complexity of the Ising model

We turn first to the case of the Ising partition function. It turns out that there is a surprising dichotomy which hinges on the structure of the underlying graph or lattice.

(6.4.1) Theorem. Computing the Ising partition function for a general graph is #P-hard even in the ferromagnetic case where each interaction energy $J_e = +1$, and there is no external magnetic field.

This follows from Theorem 6.2.9 because the Ising partition function is an evaluation of the Tutte polynomial along the hyperbola H_2. However we will give a direct proof due to Jerrum and Sinclair (1990) as it highlights a useful method;

Proof. We know from §4.3 that we can express Z in the form

$$Z = \sum_{k=0}^{|E|} b(k)^{\beta(|E|-2k)}$$

where $b(k)$ is the number of cuts of size k. Hence

$$Z = 2^{\beta|E|}p(4^{-\beta})$$

where

$$p(x) = \sum_{k=0}^{|E|} b(k)x^k.$$

Suppose the value of p is known at points $\beta = 0, 1, 2, ..., |E|$; that is at the points $x = 1, 4^{-1}, 4^{-2}, ..., 4^{-|E|}$. Then using Lagrange interpolation we could recover the coefficients $b(k)$ in polynomial time. [Note: using Newton's formula this process can be carried out using only rational arithmetic and the numerators and denominators remain polynomially bounded.] But the leading coefficient in p is twice the number of maximum cut-sets in G. And determining this is $\#P$-hard. $\qquad\square$

However, in what is probably the most difficult of the known polynomial time counting algorithms, Kastelcyn (1967) proved the following theorem.

(6.4.2) Theorem. There is a polynomial time algorithm which will determine the Ising partition function of any planar graph when there is no external field.

Because of its importance and widespread application, there is a large literature on this algorithm. It is not easy; a very good treatment may be found in Lovász-Plummer (1986 Chapter 6) and it would be pointless to repeat this here.

Computing ground states of spin systems

A *ground state* of any spin system is a state of minimum energy. Since the energy of a state σ is

$$H(\sigma) = -\Sigma J_{ij}\sigma_i\sigma_j - \Sigma M\sigma_i,$$

it is easy to see that the following is true.

(6.4.3) When $M = 0$ and $J_{ij} \geq 0$, finding a ground state is equivalent to finding an edge cut of maximum weight, and this is a known NP-hard problem.

When G is planar an edge cut in G maps into an Eulerian subgraph in the dual planar graph G^*. This brings us to the following question.

The *Chinese postman problem* is that of finding the minimum number of edges to double in a graph G in order to make it Eulerian. It can be solved in polynomial time, see Edmonds and Johnson (1973).

An easy observation is that if $q(G)$ is the minimum number of edges which have to be doubled in G in order to make G Eulerian, then $q(G)$ equals the minimum number of edges which need to be deleted from G in order to make it Eulerian. Moreover if $E' \subset E(G)$ is the set of edges of G which are doubled in an optimal Chinese postman tour, then the removal of E' from G gives a maximum cardinality Eulerian subgraph of G. This shows that finding a maximum cardinality Eulerian subgraph of a graph can be done in polynomial time.

In other words:

(6.4.4) For the ferromagnetic Ising model on a planar graph with zero external field the ground state can be found in polynomial time.

Turning now to the antiferromagnetic case, it is clear from (6.4.3) that the following holds.

(6.4.5) When there is no external magnetic field and $J_{ij} \leq 0$, then determining the ground state is the problem of finding a cut of minimum weight.

But by the max flow mincut theorem, this is well known to be in P for all graphs, not just for planar graphs.

We summarise all these observations in the following theorem.

(6.4.6) Theorem. When there is no magnetic field, finding a groundstate in the antiferromagnetic case is in P but in the ferromagnetic case the problem is NP-hard for general graphs but is in P for planar graphs.

We now turn to the case where there is an external field.

Finding the ground state of a planar Ising model in a non-zero magnetic field is clearly at least as hard as the following problem:

Instance: Planar graph $G = \langle V, E \rangle$.
Output: Minimum value of

$$\sum_{(i,j) \in E} \sigma_i \sigma_j + \sum_{i \in V} \sigma_i$$

where $\sigma_i \in \{-1, 1\}$ for each $i \in V$.

This corresponds to planar spin glass with all its interactions antiferromagnetic ($J_{ij} = -1$ for all $(i, j \in E)$) and a magnetic field $F = 1$.

We follow Barahona (1982) and reduce this to the following NP-hard problem.

Instance: Planar graph G, regular of degree 3.
Output: Maximum cardinality of a stable set.

(6.4.7) Theorem. If there is an external magnetic field, determining the ground state is NP-hard even in the planar case and where all $J = -1$.

Proof: Let $G = \langle V, E \rangle$ be the cubic planar input.

Associate a variable $x_i \in \{0, 1\}$ to each vertex $i \in V$. There is a stable set of cardinality k iff there is an assignment of values $\{x_i : i \in V\}$ satisfying

$$L = \sum_{i \in V} x_i - \sum_{(i,j) \in E} x_i x_j \geq k.$$

Setting $\sigma_i = 2x_i - 1$ we obtain

$$L = \left(-\frac{1}{4} \sum_{i \in V} \sigma_i - \frac{1}{4} \sum_{(i,j) \in E} \sigma_i \sigma_j \right) + |V| / 8.$$

Hence there exists a stable set of cardinality k in G iff there is an assignment $\{\sigma_i\}$ such that $H \leq (|V| / 2) - 4k$. $\square$

6.5   Reliability and other computations

Of the various complexity questions associated with counting those concerning reliability were among the first to be considered, see in particular Ball (1977), Valiant (1979a), Jerrum (1981) and Ball and Provan (1983).

The original proofs of hardness were based on Lagrange interpolation methods. However, in view of the fact (see §3.4) that we can represent the all terminal reliability in terms of the Tutte polynomial evaluated along $x = 1$, we have immediately from Theorem 6.2.11 the result:

(6.5.1) Determining the all terminal reliability of a graph is $\#P$-hard even if it is bipartite and planar.

A closely related quantity is the *two terminal reliability* $R(G; s, t; p)$ which represents the probability that under the same conditions, two specified vertices s and t are connected in G. This was shown to be $\#P$-complete for general graphs by Valiant (1979a), a stronger result due to Provan (1986) is:

(6.5.2) Determining the two terminal reliability of a planar graph of maximum degree 3 is $\#P$-hard.

An interesting result of Provan and Ball (1983) concerns the difficulty of finding approximations to the two terminal reliability and is stated as follows.

Given $\alpha < 1$, $0 \leq p \leq 1$, find a number r such that

$$\alpha r < R(G; s, t; p) < \frac{r}{\alpha}.$$

They show that finding such an r is $\#P$-hard.

We return to questions of approximation and in particular to approximating reliability in §8.

Another result of Provan (1986), highlights how difficult reliability calculations can be. It concerns reliability in directed networks.

(6.5.3) Theorem. Computing the source-sink reliability even in acyclic directed planar graphs of maximum degree three is $\#P$-hard.

It is clear that the computational problems involved in carrying out percolation calculations are of very much the same nature as those for reliability. The main differences are that in percolation one tends to be more concerned with regular lattices.

For general graphs, it follows immediately from our earlier Theorem 6.2.11 that almost every computation concerned with the random cluster model is likely to be hard. To see this, note that since the partition function $Z(p, Q)$ of the random cluster model is a specialisation of the Tutte polynomial along the hyperbola $H_Q \equiv (x - 1)(y - 1) = Q$, we have as a corollary of (6.2.11):

(6.5.4) Unless $Q = 1$ or 2, determining the partition function $Z(p, Q)$ is $\#P$-hard even in the bipartite planar case.

A consequence of this, observed by Annan (1993), is the following statement.

(6.5.5) In the random cluster model with $Q \neq 1$, even determining the probability that an edge is open is $\#P$-hard.

Other consquences of Theorem 6.2.11 and of some of the specialisations listed in §3.7 are the following statements.

(6.5.6) For bipartite planar graphs the problem of counting each of the following is $\#P$-hard: a) forests b) acyclic orientations c) k-colourings for $k \geq 3$.

(6.5.7) Determining the weight enumerator of a linear code is $\#P$-hard.

(6.5.8) Counting the number of regions into which hyperplanes divide $\mathbf{R}^n$ is $\#P$-hard.

Many other such results follow from applying Theorem 6.2.11 to the various other interpretations of T to be found in Brylawski and Oxley (1992).

6.6 Additional notes

Hemion (1992) gives an up-to-date account of the knot classification problem. Tables of knot polynomials of prime knots up to 10 crossings can be found in Jones (1987). An example which would show that the Jones polynomial does not characterise the unknot will have at least 14 crossings. The complexity of questions such as 6.3.11 are considered in Welsh (1992b).

An interesting collection of $\#P$-complete problems is given by Linial (1986) and a rich source of information on algorithmic questions in this area is Lovász and Plummer (1986).

Complexity questions from statistical physics are considered in Jerrum (1987) and Welsh (1990). For more on the complexity of reliability questions see Colbourne (1987). An entirely different approach to the "complexity of a knot" may be found in Soteros, Sumners and Whittington (1992).

7 The Complexity of Uniqueness and Parity

7.1 Unique solutions

Deciding whether or not a problem has a unique solution, such as whether a graph has a unique Hamilton circuit, might seem to be easier than deciding whether there is any solution. This does not seem to be the case and questions concerning uniqueness are among the most subtle in complexity theory. The two main complexity classes, US and UP, which are associated with uniqueness questions have a status in the complexity hierarchy which is far from clear, even though they have been well established now for about 10 years. The more interesting class, UP, has intimate connections with the existence of one-way functions and hence with cryptography, however as we shall see, it is almost totally devoid of known natural members.

Blass and Gurevich (1982) introduced the following class of languages which they called *UNIQUE SOLUTION*, abbreviated to US. It is defined as follows:

(7.1.1) $L \in US \iff \exists$ a polynomially bounded polynomial time relation $R : \Sigma^* \times \Sigma^* \to \{0, 1\}$ such that $x \in L \iff \exists$ *unique* y satisfying $R(x, y)$.

In other words;

(7.1.2) $L \in US \iff \exists$ a polynomial-time ndtm M such that $x \in L$ if and only if M, on input x, has exactly one accepting computation path.

For example consider the problem:

UNIQUE-SAT
Instance: Boolean formula F in conjunctive normal form.
Answer: "Yes" if F has exactly one satisfying assignment.

Now recall the fundamental NP-complete problem $TM\text{-}COMP$ defined in (1.3). Since the reduction $TM\text{-}COMP \propto SAT$ given by Cook is parsimonious, it follows immediately that US has complete members and moreover:

(7.1.3) *UNIQUE-SAT* is complete for US.

Proof: Take the relation $R(x, y)$ to be "x is a Boolean formula and y is a truth assignment satisfying it". $\square$

Similarly we see that if A is any NP complete language and $UNIQUE-A$ is the associated uniqueness question, then if there exists a parsimonious reduction between A and SAT, then the corresponding uniqueness question is complete for US. Thus for example:

(7.1.4) *UNIQUE HAMILTONIAN* is complete for US.

(7.1.5) For $k \geq 3$, *UNIQUE k-COLOURABILITY* is complete for US.

A fundamental observation is the following due to Blass and Gurevich (1982).

(7.1.6) Proposition *UNIQUE-SAT* is NP hard.

Proof: Let $E = E(x_1, \ldots, x_n)$ be a Boolean expression in conjunctive normal form $C_1 \wedge \ldots \wedge C_m$.
Define $f(E)$ as follows:
If $E(1,1,...,1) = 1$ then define $f(E)$ to be any non-satisfiable expression.
If $E(1,1,...,1) = 0$ introduce a new variable y and set

$$f(E) = (C_1 \vee y) \wedge (C_2 \vee y) \wedge \ldots \wedge (C_m \vee y) \wedge (x_1 \vee \bar{y}) \wedge \ldots \wedge (x_n \vee \bar{y}).$$

$f(E)$ has a solution $x_1 = 1$, $x_2 = 1, ..., x_N = 1$, $y = 1$ and this is the only possible solution with $y = 1$. Hence, # {solutions of $f(E)$} = #{solutions of E} + 1, since any solution **x** of E gives a solution $(\mathbf{x}, 0)$ of $f(E)$, and conversely any solution $(\mathbf{x}, 0)$ of $f(E)$ gives rise to a solution of E.
Thus E is not satisfiable iff $f(E)$ is uniquely satisfiable. □
Now let us consider the place of US in the complexity hierarchy. First, note that it is easy to modify the last proof to get:

(7.1.7) co-$NP \subseteq US$.

Proof: Let $L \in$ co-NP. Then $x \in L \longleftrightarrow x$ has no solution. Let g be a reduction of $\Sigma^* \backslash L$ to SAT so that $x \in \Sigma^* \backslash L \leftrightarrow g(x)$ is satisfiable. Now consider the function f defined in the proof of (7.1.6): $g(x)$ is a Boolean expression and is not satisfiable if and only if $f(g(x))$ is uniquely satisfiable. Since $x \to f(g(x))$ is polynomially bounded, this shows that $L \in US$. □

The difference class D^p
Next we compare US with the class D^p introduced in Papadimitriou and Yannakakis (1984). D^p is an abbreviation for DIF^p and is defined by

$$\begin{aligned} DIF^p &= \{L_1 \backslash L_2 : L_1, L_2 \in NP\} \\ &= \{L_1 \cap L_2 : L_1 \in NP, L_2 \in \text{co-}NP\}. \end{aligned}$$

Example: $UNIQUE-SAT \in D^p$.

Proof: Take $L_i = \{x : x$ is a Boolean formula with at least i distinct satisfying assignments$\}$. For each fixed i, $L_i \in NP$, and $UNIQUE\text{-}SAT$ $= L_1 \cap (\Sigma^* \backslash L_2)$.   □

A modification of this idea shows

(7.1.8) $US \subseteq D^p$.

The 'obvious' complete problem for D^p is:

SAT-UNSAT
Instance: Two formulae E_1, E_2 in conjunctive normal form.
Question: Is E_1 satisfiable and E_2 not satisfiable?

Equivalently, take any NP-complete language such as $3\text{-}COLOURABIL\text{-}ITY$ and define the following problem:

3-COLOURABILITY - NON 3-COLOURABILITY
Instance: Two graphs G, H.
Question: Is G 3-colourable and H not 3-colourable?

This, and any similarly defined problem is complete for D^p.
Other problems known to be in D^p and not thought to be in NP are:

EXACT CLIQUE
Input: Graph G and integer k.
Question: Is the maximum clique in G of size exactly k?

CRITICAL NONHAMILTONIAN
Input: Graph G.
Question: Is every single edge extension of G Hamiltonian?

An intriguing question is the following:

(7.1.9) *Open Problem.* Is $NP \subseteq US$?

In other words;

(7.1.10) If $L \in NP$, does there exist a polynomial time nondeterministic machine M such that

$$x \in L \Longleftrightarrow acc_M(x) = 1?$$

This seems unlikely, as do the equivalences shown in the following proposition.

(7.1.11) Proposition. The following statements are equivalent

   (a) $NP \subseteq US$
   (b) $US = D^p$
   (c) $UNIQUE\text{-}SAT$ is complete for D^p.

The proof is easy.

7.2   Unambiguous machines and one-way functions

A nondeterministic Turing machine is *unambiguous* if no input has more than one accepting computation.

UP is the class of languages which are accepted by unambiguous ndtms which are polynomial time bounded. Clearly every deterministic machine is unambiguous and thus an immediate consequence is:

(7.2.1) $P \subseteq UP \subseteq NP$.

The importance of UP is due to the following theorem which relates it with functions which are easy to compute but hard to invert, the so called one-way functions.

(7.2.2) Theorem. One-way functions exist if and only if $P \neq UP$.

Loosely speaking, f is one-way if it is computable in polynomial time but it is difficult to solve equations of type $f(x) = y$ for given values of y. This may be for the rather trivial reason that $|x|$ is very much larger than $|y|$. Thus we are led to the notion of defining f as *honest* if there is a polynomial q such that for all $y \in$ range (f) there exists $x \in$ dom (f) such that $f(x) = y$ and $|x| \leq q(|y|)$.

A function f is said to be *one-way* if it is computable in polynomial time, is one-one, is honest and f^{-1} is not computable in polynomial time.

The relation between UP and US is subtle. UP is the class of problems (languages) for which there exists *some* NP-machine on which they have unique witnesses. That is $L \in UP$ iff there exists a polynomial time ndtm M accepting L and for *all* inputs x, $acc_M(x) \leq 1$. Hemachandra (1987) calls these machines *categorical*.

Consider for example, the following problem:
DISCRETE LOGARITHM
Instance: A prime integer p, a primitive root a modulo p, and an integer y, $0 < y < p$.
Output: The discrete logarithm of y with respect to p and a. That is, the unique integer x, $0 \leq x < p$, such that $a^x = y \bmod p$.

As presented, DISCRETE LOGARITHM is not a recognition problem and cannot possibly be a candidate for UP. However if we extend the notion of UP to include functions, so that FUP denotes the class of partial functions computable by polynomial time unambiguous Turing machines, it becomes a likely candidate for membership of $FUP \backslash FP$. This is because in a cryptographic context it is regarded as one of the stronger problems on which to base a cryptosystem, see for example Welsh (1988).

As pointed out by Johnson (1985), this problem is almost in the class FUP. However, as it stands, the domain of the function is improper. There is no known method of "uniquely verifying" either that p is prime or that a is primitive mod p. However Johnson (1990) claims that the following is an example of a function belonging to *FUP* and not known to be in *FP*.

AUGMENTED DISCRETE LOGARITHM
Input: p, a, y as above, together with short proofs that p is prime and that a is primitive mod p.
Output: The discrete logarithm of y with respect to p and a.

By the result of Pratt (1975) there exist succinct certificates for primality, and Pratt's techniques also enable one to verify primitivity in polynomial time. Thus, we know that the augmented version above belongs to *FUP*. However, showing it did not belong to *FP* would show $NP \neq P$.

7.3   The Valiant-Vazirani theorem

The following result of Valiant and Vazirani (1986) is probably the most important result in the theory of uniqueness.

(7.3.1) Theorem Let A be any *NP*-complete problem to which *SATISFI-ABILITY* is parsimoniously reducible. Let $\#A(x)$ denote the number of solutions to instance x. For each Boolean predicate Q, define the problem UA_Q by

$$UA_Q(x) = \begin{cases} 0 & \text{if } \#A(x) = 0 \\ 1 & \text{if } \#A(x) = 1 \\ Q(x) & \text{if } \#A(x) > 1. \end{cases}$$

If $UA_Q \in RP$ for any Q, then $NP = RP$.

The theorem can be better appreciated when stated more loosely in the following form.

(7.3.2) Theorem. Suppose that A is parsimoniously reducible to *SAT*, and that there exists a polynomial time algorithm which on all instances of A having 0 or 1 solutions give the correct answer, but is completely arbitrary on all other inputs. Then $NP = RP$.

First we briefly review the notion of a randomised reduction. We say that a problem A is reducible to B by a *randomised polynomial-time reduction* if there exists a probabilistic (coin flipping) polynomial time Turing machine T and a polynomial p such that $\forall x \in \Sigma^*$,

(i) if $x \notin A$ then $T(x) \notin B$

(ii) if $x \in A$ then $T(x) \in B$ with probability at least $\frac{1}{p(|x|)}$.

Note. The standard definition is to require $1/2$ instead of $1/p(|x|)$ in (ii) but it is a straightforward piece of probability to see that (ii) as given is sufficient.

It is easy to prove the following:

(7.3.3) Proposition. If A is reducible to B by a randomised polynomial time reduction and $B \in RP$ then $A \in RP$.

(7.3.4) Corollary. If an NP complete problem A is reducible to B by a randomised polynomial time reduction and $B \in RP$ then $NP = RP$.

Proof of the Valiant-Vazirani theorem.

We will regard truth assignments to the variables $x_1, ..., x_m$ as vectors from the vector space F_2^m. For $u, v \in F_2^m$ we denote by $u.v$ their inner product over F_2.

First we observe:

(7.3.5) Lemma. If f is a CNF formula in $x_1, ..., x_m$ and $w_1, ..., w_k \in (F_2)^m$ then one can construct in linear time a formula f_k' whose solutions v satisfy E and also the k equations

$$v.w_1 = v.w_2 = ... = v.w_k = 0.$$

Furthermore, one can construct a polynomial-size CNF formula f_k in variables $x_1, ..., x_m, y_1, ..., y_{m'}$ for some m' such that there is a bijection between solutions of f_k and f_k', defined by equality on the $x_1, ..., x_m$ values.

Proof. It is clearly sufficient to prove the lemma for $k = 1$. Then

$$f' \equiv f_1 \wedge (x_{i_1} \oplus x_{i_2} \oplus ... \oplus x_{i_j} \oplus 1),$$

where $\oplus$ is exclusive-or, and $i_1, i_2, ..., i_j$ are the indices of those x_i that have the value 1 in w. Also f_1 is the CNF equivalent of the formula

$$f \wedge (y_1 \Leftrightarrow x_{i_1} \oplus x_{i_2}) \wedge (y_2 \leftrightarrow y_1 \oplus x_{i_3}) \wedge ...$$
$$\wedge ... \wedge (y_{j-1} \Leftrightarrow y_{j-2} \oplus x_{ij}) \wedge (y_{j-1} \oplus 1). \qquad \square$$

The key to the proof of the main theorem is the following idea.

Let S be an arbitrary non empty subset of $\{0,1\}^n$. Let $w_1, ..., w_n$ be independently chosen random vectors from $\{0,1\}^n$.

Let

$$H_i = \{w : v.w_i = 0\}$$

so that H_i is a hyperplane in F_2^n. Define the nested sequence $S_0 \supset S_1 \supset S_2 ... \supseteq S_n$ by,

$$S_0 = S, \quad S_1 = S_0 \cap H_1, \quad S_2 = S_1 \cap H_2, ...$$

Then the claim is the following:

(7.3.6) Lemma. The probability that for some i, $|S_i| = 1$, is at least $1/2$.

There exist several purely combinatorial proofs of this. Two, due to M. Jerrum and M. Rabin may be found in the paper of Valiant and Vazirani.

So how do we use this randomised reduction from SAT to $USAT_Q$?

We suppose $\mathcal{A}$ is our algorithm which works for instances of $USAT_Q$.

Suppose we are given an instance f of SAT in variables $x_1, ..., x_m$. Our aim is to construct a new instance g such that:

(i) if f is satisfiable then there is a "reasonable probability" that g has *exactly* one satisfying assignment,

(ii) if f is non-satisfiable then g is not satisfiable.

We now use these lemmas to prove Theorem 7.3.1. Given an instance f of SAT, in variables $x_1, ..., x_m$, randomly choose k from $\{1, ..., m\}$, randomly choose k vectors $w_1, ..., w_k \in \{0, 1\}^m$ and construct f_k as above. Then from (7.3.6) it is easy to prove:

(7.3.7) $Pr\{f_k \text{ is satisfiable} \mid f \text{ is satisfiable}\} \geq 1/2\,m$.

Also since f_k is not satisfiable if f is not, we have the required randomised reduction. □

We now show some applications of the above ideas.

It is natural to conceive of situations where, in building an algorithm, a major part of the difficulty is deciding what to do when confronted with choice. This line of thought might suggest that for problem instances which are guaranteed to have exactly one solution there might be fast algorithms for *finding* that solution, even though the general problem remains hard. Steepest descent and greedy algorithms are two common techniques in optimisation which come to mind and have this flavour. However this does not seem to be the general situation as we now illustrate.

Suppose that $\mathcal{A}$ is a polynomial time algorithm which given as input a uniquely 3-colourable graph will find its 3-colouring.

We define the function UA as follows. For any input x, apply $\mathcal{A}$ and if it produces an 3-colouring check that it is correct, clearly this checking process can be done in polynomial time. Define

$$UA(x) = \begin{cases} 0 & \text{if } \mathcal{A}(x) \text{ does not produce a 3-colouring} \\ 1 & \text{if } \mathcal{A}(x) \text{ produces a correct 3- colouring.} \end{cases}$$

Then $UA(x)$ satisfies the conditions of the Valiant-Vazirani theorem. Since by Berman and Hartmanis (1977), 3-$COLOURING$ is parsimoniously reducible to SAT we have proved:

(7.3.8) If there exists a polynomial time algorithm which will find a 3-colouring of a uniquely 3-colourable graph then $NP = RP$.

Now the role of 3-colouring in the above argument is one of motivation. The argument clearly extends to more general situations and proves the following useful result.

(7.3.9) Theorem. For any NP language which is parsimoniously reducible to SAT, finding a solution, even for those inputs which are guaranteed to have exactly one solution, can be done in polynomial time only if $NP = RP$.

Since this is regarded as most unlikely, what we are saying is that knowing there is exactly one solution is not much help.

7.4   Hard counting problems not parsimonious with SAT

Recall that a *parsimonious transformation* between a problem X and a problem Y is a polynomial time transformation f such that if $\#(X, x)$ is the number of solutions of problem X with input x, then $\#(X, x) = \#(Y, f(x))$.

It appears to be the case that parsimonious transformations can be found between most of the "classical " NP-complete problems. However, there can be no hope of a general result of this nature, since as we now show there is also a plethora of NP-complete problems between which no parsimonious transformations can exist.

Consider first the following, somewhat artificial, example.

DOUBLE-SAT
Instance: Boolean formula F in conjunctive normal form with an additional clause of the form $\{y \vee \bar{y}\}$ where y is a variable which does not appear in any other clause.
Question: Is F satisfiable?

It is easy to see that *DOUBLE-SAT* is NP-complete, since it is trivially in NP and obviously SAT reduces to it. However, because of the additional clause, we see that for any instance F, of *DOUBLE-SAT*, the number of satisfying instances is even. But since SAT itself can have instances with an odd number of satisfying assignments, we have shown:

(7.4.1) *DOUBLE-SAT* is an NP-complete problem which is not parsimoniously reducible to SAT.

Nevertheless, despite the absence of a parsimonious transformation we see:

(7.4.2) Counting solutions to *DOUBLE-SAT* is $\#P$-complete.

Proof: Given any instance X of SAT we may transform X to $\phi(X)$ by

$$\phi(X) = X \wedge \{y \vee \bar{y}\}$$

where y is not one of the variables of X. Now observe that $\phi(X)$ is a valid input to *DOUBLE-SAT* and $\phi(X)$ has exactly twice the number of solutions as X, so that *#DOUBLE-SAT* is *#P*-hard. But *#DOUBLE-SAT* clearly belongs to *#P* so it must be *#P*-complete. □

A more natural problem which is *NP*-complete but for which the natural witness set is not parsimoniously transformable to *SAT* is the following

CHROMATIC INDEX
Instance: A k-regular graph G.
Question: Does there exist a k-colouring of the edge set such that no two incident edges are the same colour?

The 'natural witness' of such a k-colouring is a partition $(E_1, ..., E_k)$ of the edge set such that for each i, no two members of E_i share a common vertex.

Holyer (1981) proved that when $k = 3$ this problem is *NP*-complete and Leven and Galil (1983) showed that this result holds for general k. However as pointed out in Edwards and Welsh (1983), the associated counting problem cannot be parsimoniously transformable to *SAT*. The reason for this is that the following problem is in *P*.

UNIQUE EDGE COLOURING
Instance: A k-regular graph G ($k \geq 4$).
Question: Does G have a unique edge colouring?

There is a linear time algorithm for this problem because of the following theorem of Thomason (1978) which states that for $k \geq 4$, the only graphs which have a unique edge colouring are the stars $K_{1,k}$.

Hence, if there were a parsimonious transformation from *SAT* to *CHROMATIC INDEX* we could use this to obtain a polynomial time algorithm for *UNIQUE-SAT*. Thus we would have shown that a complete problem for *US* was in *P* and would have shown *NP = P*. To sum up:

(7.4.3) There is no parsimonious transformation from *SAT* to *CHROMATIC INDEX* unless *NP = P*.

7.5 The curiosity of parity

There is no obvious relationship between the difficulty of counting the members of a set and determining its parity, other than the obvious remark, that if it is easy to count then it is easy to decide parity.

This is typified by the following statements.

(7.5.1) There is no known subexponential time algorithm which will decide whether the number of Hamilton circuits of a graph is even or odd.

(7.5.2) Although computing permanent is *#P*-hard there is a polynomial time algorithm which will decide whether it is even or odd.

The study of the complexity of questions concerning parity originated in the paper of Papadimitriou and Zachos (1983), and has had great impetus in the last few years due to the striking results of Toda (1989) which shows that a parity oracle together with randomness is extremely powerful, and at least as strong as any problem in the polynomial hierarchy.

PARITY-P, denoted $\oplus P$, is the class of languages L such that there exists a polynomial time ndtm M with the property that

$$x \in L \Leftrightarrow \mathrm{acc}\,(M, x) \text{ is odd.}$$

As a class of languages, $\oplus P$ seems difficult to analyse. It is not difficult to show:

(7.5.3) $L \in \oplus P \Leftrightarrow \Sigma^* \backslash L \in \oplus P.$

Proof: By definition there exists a relation R such that

$$x \in L \Leftrightarrow |\{y : R(x,y) = 1\}| \text{ is odd.}$$

Consider the relation $\bar{R}$ defined by

$$\bar{R}(x,y) = \begin{cases} 1 & \text{if} \quad R(x,y) = 0 \\ 0 & \text{if} \quad R(x,y) = 1. \end{cases}$$

Then $\left|\{y : \bar{R}(x,u) = 1\}\right| = 2^{p(x)} - |\{y : R(x,y) = 1\}|$ for some polynomial p and this proves (7.5.4).

Alternatively, and this may be easier to visualise; $L \in \oplus P$ if and only if there exists a polynomial time ndtm M accepting L in which the number of accepting leaves is odd. Change M to $\bar{M}$ so that $\bar{M}$ accepts if M doesn't and count the number of accepting configurations of $\bar{M}$.   □

Since UP is the class of languages accepted by polynomial time ndtms with exactly one accepting computation it is trivial that $UP \subseteq \oplus P$. There is however a stronger result which relates UP to the slightly strange class of languages, *Few P*, introduced by Allender (1985).

Few P is a natural generalisation of UP and consists of all languages recognised by polynomial time ndtms for which the number of accepting computations is bounded by a fixed polynomial in the size of the input.

Again, almost by definition, we have

(7.5.4) $UP \subseteq Few P \subseteq NP.$

We can now prove a recent theorem of Cai and Hemachandra (1989). Its proof is elegant and gives a good insight into the character of $\oplus P$.

(7.5.5) Theorem. *Few P $\subseteq \oplus P$.*

Proof: Let M be a ndtm which witnesses that $A \in$ *Few P*. Then, by definition, there exists a polynomial p, such that for all x,

$$A = L(M) \quad \text{and} \quad acc_M(x) \leq p(|x|).$$

Construct M', another ndtm as follows:
input x;
guess an integer k: $1 \leq k \leq p(|x|)$;
guess a sequence $y_1, ..., y_k$ of computation paths for M on x such that $y_1 < ... < y_k$ in the obvious lexicographic ordering.
In M on x with path y_i is accepting for all i, $1 \leq i \leq k$, then M' halts accepting. Else M' rejects.

Thus M' has zero accepting configurations if $x \notin A$, but if $x \in A$, then if M on x has m accepting computations then M' has

$$\sum_{k=1}^{m} \binom{m}{k} = 2^m - 1$$

accepting computations, which is an odd number. Also since $m \leq p(|x|)$, M' works in polynomial time. Hence $A \in \oplus P$. $\square$

The relationship of $\oplus P$ to NP is not clear. However like NP and $\#P$, $\oplus P$ has complete members, and the proof of this parallels our earlier proof that $\#P$ has complete members.

To see this, define $\oplus SAT$ to consist of those conjunctive normal forms which have an odd number of satisfying assignments.

(7.5.6) *PARITY-P* has a complete language $\oplus SAT$, and also the parity version of any NP-complete language which is parsimoniously reducible to *SAT*.

Proof: Consider the generic problem *TM-COMP*. Let $\oplus TM$-*COMP* consist of those instances with an odd number of accepting computations. Parsimonious reductions clearly preserve parity. $\square$

Another way of thinking about NP, $coNP$ and $\oplus P$ is as follows.

Suppose that we regard ndtms as ordinary (deterministic) Turing machines with an additional "free gate" $\{\vee\}$.

Counting machines, can be regarded as extended Turing machines with domain the natural numbers and $\{+\}$.

Co-nondeterministic machines are Turing machines with an additional free gate $\{\wedge\}$.

In a similar spirit, *Parity machines* have an additional gate $\{\oplus\}$.

All these machines accept if the final state is 1.

Goldschlager and Parberry (1986) study $\oplus P$ in connection with circuit complexity and asked the following question which remains unanswered.

(7.5.7) Problem. Does $NP \cap coNP \cap \oplus P = P$?

Goldschlager and Parberry also ask whether there exists a "natural problem" which is $\oplus P$-complete? By implication, $\oplus SAT$ is not regarded as natural and the real question here must be whether there exists a problem which is complete for $\oplus P$ but which is not the parity version of some standard NP-complete problem.

Arithmetic mod k

The Valiant-Vazirani theorem about unique solutions can also be exploited to deal with counting mod k. The argument is very similar; suppose that $\mathcal{A}^k$ is an algorithm which, given a Boolean formulae f, answers as follows:

$$\mathcal{A}^k(f) = \begin{cases} 0 & \text{if } \#f = 0 \bmod k, \\ 1 & \text{if } \#f = 1 \bmod k, \\ Q(f) & \text{otherwise.} \end{cases}$$

Then $\mathcal{A}^k$ will certainly answer correctly for those instances of SAT in which there are exactly zero or 1 satisfying assignments. Hence by Theorem 7.3.1 we have proved:

(7.5.8) If A is any NP-complete language parsimoniously reducible to SAT then deciding whether $\#A(x) = 0$ or 1 mod k is NP hard unless $RP = NP$.

However it is not only NP-hard problems which give rise to hard parity problems, as we now show.

(7.5.9) If there exists a polynomial time algorithm for computing permanent mod k for any k which is not a power of 2, then $NP = RP$.

Proof. In his proof that computing the permanent is $\#P$ hard, Valiant (1979b) gives a transformation from a SAT formula f to a $(0,1)$-matrix B such that $\text{perm}(B) = \#SAT(f).4^t$ where $t = t(f)$ is polynomial-time computable. Thus provided k is not a power of 2; $\#SAT(f) = 0 \bmod k \Leftrightarrow \text{perm } B = 0 \bmod k$.

Thus, suppose $\mathcal{A}$ is a polynomial time algorithm for permanent mod k. Then we could use the above transformation combined with $\mathcal{A}$ to give a randomised polynomial time algorithm which answered correctly for all instances of SAT having exactly zero or one solution. But by Theorem 7.3.1, this would imply $NP = RP$. $\qquad\square$

This is an example of a transformation which is not parsimonious but it is weakly parsimonious in the sense of Johnson (1990). A reduction f between problems A, B is called *weakly parsimonious* if f maps instances of A to instances of B in such a way that if x is any input of A, the problem of computing $\#(A, x)$ given $\#(B, f(x))$ can be done in polynomial time.

What makes parity questions difficult to analyse is that there are a surprisingly large number of questions where although exact counting is $\#P$-hard, determining the parity is easy, that is polynomial time computable.

A large class of such examples follows from the fact that although Tutte invariants are, apart from the special points, #P-hard to evaluate, their parity is easy to decide. More precisely we can prove the following result.

(7.5.10) Theorem. For integers a, b and G any graph, the parity of $T(G; a, b)$ can be calculated in polynomial time.

Proof. Consider the representation

$$T(G; a, b) = \Sigma t_{ij} a^i b^j,$$

where from (3.2.9) we know that the t_{ij} are nonnegative integers. Then it is an easy exercise to check that if $a' = a \pmod 2$ and $b' = b \pmod 2$ then

(7.5.11) $$T(G; a, b) \bmod 2 = T(G; a', b') \bmod 2.$$

But depending on whether (a', b') equals (1,1) (1,0) or (0,1) the right hand side of (7.5.11) is given by $T(G; a'', b'')$ where (a'', b'') equals (1,1),(- 1,0) and (0,-1). But at each of these points T can be evaluated exactly in polynomial time. □

7.6   Toda's theorem on parity

We now turn to a major result linking $\oplus P$ with other more well known complexity classes. Loosely speaking it says that an oracle which will decide parity combined with an oracle which will produce random bits is at least as powerful as an oracle for anything belonging to the polynomial hierarchy.

In order to state the result precisely we need to introduce the following operators $\oplus$ and BP.

(7.6.1) Definition. For any class **C** of sets $\oplus.\mathbf{C}$ is the class of sets L such that for some set $A \in \mathbf{C}$, some polynomial p, and all $x \in \Sigma^*$,

$$x \in L \Leftrightarrow \left| \{ w \in \{0, 1\}^{p(|x|)} : x\#w \in A \} \right| \quad \text{is odd.}$$

Here $x\#w$ denotes the string $x \in \Sigma^*$ separated from the string $w \in \Sigma^*$ by the special symbol $\#$.

Thus for example $L \in \oplus.P$ iff there exists some p-relation R such that

$$x \in L \Leftrightarrow |\{y : R(x, y)\}| \quad \text{is odd.}$$

It is therefore clear that,

(7.6.2) $$\oplus.P = \oplus P,$$

and we can therefore use both definitions of *PARITY-P* interchangeably.

(7.6.3) Definition. For any class **C** of sets, $BP.\mathbf{C}$ is the class of sets L such that for some set $A \in \mathbf{C}$, some polynomial p, some $\epsilon > 0$, and all $x \in \Sigma^*$,

$$x \in L \Rightarrow Pr\{w \in \{0,1\}^{p(|x|)} : x\#w \in A\} \geq \frac{1}{2} + \epsilon$$

$$x \notin L \Rightarrow Pr\{w \in \{0,1\}^{p(|x|)} : x\#w \in A\} \leq \frac{1}{2} - \epsilon.$$

Again it is fairly obvious that the operator BP applied to P gives the more familiar class BPP of randomised algorithm theory.

Toda's theorem is the following striking result.

(7.6.4) Theorem. The polynomial hierarchy PH is contained in the class $BP. \oplus P$.

As an exercise in assessing the worth of this theorem, consider the question of how to use a $\oplus P$ oracle, but no randomness, to settle an NP-complete problem. It is by no means clear that this can be done. However, with the help of a source of randomness it, and a lot more, can be done.

7.7 Additional notes

A very good survey of problems connected with questions of uniqueness is given by Johnson (1990). Grollmann and Selman (1984) point out the relation between one way functions and cryptography. A recent survey is Selman (1992). The survey of Schöning (1990) is an invaluable source for structural questions.

Theorem 7.5.10 gives many examples of hard counting problems whose parity version is easy. However Annan (1993) has shown that computing Tutte invariants mod p for $p \neq 2$ can be NP hard or in FP.

For further results related to Toda's theorem (7.6.4) see Beigel (1990), Beigel, Gill and Hertrampf (1990), Tarui (1991) and Toda and Ogiwara (1992).

8 Approximation and Randomisation

8.1 Metropolis methods

Two classical applications of Monte Carlo methods are the evaluation of high dimensional integrals and sampling the state space of a Markov chain, using the so called "Metropolis method".

Here we shall be concerned with both of these problems, and more particularly with questions of rates of convergence.

The basic idea is to construct an irreducible finite Markov chain whose states are configurations on which we simulate a random walk. More specifically, suppose that we are given some strictly positive probability measure π on a finite set Ω and we wish to generate this measure. The idea is to construct an irreducible Markov chain with transition matrix P and state space Ω and which has π as its stationary distribution. Carrying out a random walk on this chain constitutes a randomised algorithm for generating π.

In order to construct such a chain we use the following very useful property of Markov chains. This is, that a sufficient condition that the irreducible stochastic matrix P has π as its stationary distribution, is that P satisfies the *detailed balance* condition for π, namely:

(8.1.1) For each pair of states $i, j \in \Omega$,

$$\pi(i)P(i,j) = \pi(j)P(j,i).$$

To construct P we proceed as follows. Start with any irreducible stochastic matrix $A = A(i,j)$ having state space Ω. Let $G : \mathbf{R}^+ \to [0,1]$ be any function satisfying $G(x) = xG(x^{-1})$. Define the matrix $Q(i,j)$, $i,j \in \Omega$, by

$$Q(i,j) = G\left(\frac{\pi(j)A(j,i)}{\pi(i)A(i,j)}\right).$$

Now modify A to P by

$$
\begin{aligned}
P(i,j) &= A(i,j)Q(i,j) \qquad i \neq j, \\
P(i,i) &= 1 - \sum_{j \neq i} P(i,j).
\end{aligned}
$$

It is easy to verify that we are choosing Q so that the resulting matrix P satisfies the detailed balance condition for the distribution π.

As for finding $G : \mathbf{R}^+ \to [0, 1]$ satisfying $G(x) = xG(x^{-1})$, this is easy. Two of the many solutions are

$$G(x) = \min(x, 1),$$

(this is the one usually called the Metropolis method) and

$$G(x) = \frac{x}{1 + x}.$$

(8.1.2) Example: Suppose we are aiming to simulate the ferromagnetic Potts model by this method. We know from §4.4 that the stationary equilibrium distribution we wish to achieve is

$$\pi(\sigma) = e^{-\beta H(\sigma)}/Z.$$

Choose our initial matrix to be the transition matrix which allows transitions between any pair of neighbouring states $\sigma \rightsquigarrow \sigma'$, where states are *neighbours* if their spins differ in one place, and all such neighbours are equally likely. Then take $G(x) = \min(1, x)$ and it is straightforward to check that the Metropolis algorithm is the following.

(i) If $H(\sigma') \leq H(\sigma)$ then move to σ'.

(ii) If $H(\sigma') > H(\sigma)$ then move to σ' with probability $e^{-\beta(H(\sigma') - H(\sigma))}$.

This algorithm is essentially the *single spin-flip* process. Notice that because σ' and σ differ only in one site it is easy to implement and the rules can be streamlined to

(iii) Choose site i at random, and state $X \in \{1, ..., Q\}$ at random and change colour of i to X *either* if it does not increase the energy H *or* if it does, do so with probability $e^{-\beta \Delta(H)}$ where $\Delta(H)$ is the resulting increase in H. $\qquad \square$

(8.1.3) Example. Random cluster model. Recall that in this system we have a Gibbs probability on the subsets of edges of a graph G given by

$$P(A) \propto (p/q)^{|A|} Q^{-r(A)} \qquad A \subseteq E.$$

Define an initial transition matrix which has states consisting of all edge sets $A \subseteq E(G)$, and in which transitions from A are uniformly distributed among the neighbours of A. Here neighbour will mean any set A' such that A, A' differ in just one edge. Then it is straightforward to check that the standard Metropolis algorithm consists of:

(i) if $\Delta = (p/q)^{|A'| \setminus |A|} Q^{r(A) - r(A')} \geq 1$ then move to A';

(ii) if $\Delta < 1$, move to A' with probability Δ but otherwise stay at A. $\quad \square$

In both of the above examples the method converges to the right answer. However, there is no indication of the rate of convergence and it is this that is important. We return to this in §8.4.

8.2   Approximating to within a ratio

We know that computing the number of 3-colourings of a graph G is $\#P$-hard. It is natural therefore to ask how well can we approximate it?

For positive numbers a and $r \geq 1$, we say that a third quantity $\hat{a}$ *approximates a within ratio r* or is an *r-approximation* to a, if

$$(8.2.1) \qquad\qquad r^{-1}a \leq \hat{a} \leq ra.$$

In other words the ratio $\hat{a}/a$ lies in $[r^{-1}, r]$.

Now consider what it would mean to be able to find a polynomial time algorithm which gave an approximation within r to the number of 3-colourings of a graph. We would clearly have a polynomial time algorithm which would decide whether or not a graph is 3-colourable. But this is NP-hard. Thus no such algorithm can exist unless $NP = P$.

But we have just used 3-colouring as a typical example and the same argument can be applied to any function which counts objects whose existence is NP-hard to decide. In other words:

(8.2.2) Proposition. If $f : \Sigma^* \to \mathbf{N}$ is such that it is NP-hard to decide whether $f(x)$ is non-zero, then for any constant r there cannot exist a polynomial time r-approximation to f unless $NP = P$.

However, in an important paper, Stockmeyer (1985) proves the following result, which, in a sense, complements Toda's theorem (1.8.5) relating $\#P$ with the polynomial hierarchy.

(8.2.3) Theorem. Let $f \in \#P$, and let d be a positive integer. Then there exists a deterministic Turing machine M, which if equipped with a Σ_2^p-oracle, will, on input $\langle x, \epsilon \rangle$, output $g(x)$ which approximates $f(x)$ to within ratio $(1 + \epsilon |x|^{-d})$. Moreover its running time is bounded by a polynomial in $|x|$ and ϵ^{-1}.

We now turn to consider a randomised approach to counting problems and make the following definition.

An *ϵ-δ-approximation scheme* for a counting problem f is a Monte Carlo algorithm which on every input $\langle x, \epsilon, \delta \rangle$, $\epsilon > 0$, $\delta > 0$, outputs a number $\tilde{Y}$ such that

$$Pr\{(1 - e)f(x) \leq \tilde{Y} \leq (1 + \epsilon)f(x)\} \geq 1 - \delta.$$

It is important to emphasise that there is no mention of running time in this definition but of course, ideally, one would like an algorithm which ran in time which is a polynomial function of $|x|$, ϵ^{-1} and δ^{-1}.

The origin of this idea of approximation in counting problems seems to be in the papers of Karp and Luby (1983) and Stockmeyer (1985), though the idea of approximating objects such as the chromatic number goes back much further.

Karp and Luby consider in some detail the DNF-counting problem. This is the classic problem of counting the number of satisfying assignments of a propositional formula F which is the disjunction of m clauses $C_1, ..., C_m$. Each clause is a subset of literals from the n Boolean variables $X_1, ..., X_n$.

We know from §1.4 that this is a $\#P$-complete problem. It is useful first to consider the crude approach which amounts to little more than sampling from a finite population. Let U be the set of 2^n different assignments to the X_i, regard F as a function $F : U \rightarrow \{0, 1\}$ and our goal is to estimate the size of the "good" set $G \subseteq U$, consisting of those $u \in U$ for which $F(u) = 1$.

The Monte Carlo algorithm consists of selecting N random elements u_i from U, setting $Y_i = 1$ if $F(u_i) = 1$ and 0 otherwise and letting

$$\hat{Y} = \Sigma Y_i / N.$$

Clearly $E(\hat{Y}) = \mu = |G| / |U|$ and as $N \rightarrow \infty$, $\hat{Y}$ will converge almost surely to μ. However straightforward probabilistic inequalities show that the following are true.

(8.2.4) Provided $\epsilon \leq 2$, and $N \geq 4ln(2/\delta)/\mu\epsilon^2$ the above algorithm is an ϵ-δ approximation scheme.

Thus in order for this approach to work in the sense of producing a polynomial time algorithm we need

$$4ln\left(\frac{2}{\delta}\right) \leq \mu\epsilon^2 \text{poly}(|F|)$$

where $|F|$ denotes the length of the formula F. But $\mu = |G| / |U|$ and thus this amounts to

$$|U| / |G| \leq \text{poly}(|F|).$$

Note that this is what we would expect; it demands that the number of successes in sampling from U is not too small.

In general, obtaining a polynomial bound on $|U| / |G|$ presents an insurmountable barrier - not least because it will not exist.

However, using a more ingenious method, Karp, Luby and Madras (1989) construct an ϵ-δ approximation scheme for the DNF-counting problem which has running time which is bounded by

$$O(nm\epsilon^{-2}ln(2/\delta)).$$

They apply this to problems such as the network reliability problem discussed earlier, however the transformation from two terminal network reliability to DNF is such that the number of clauses in the DNF formula is the number of cut sets separating 2 vertices. This is usually exponentially large, so this approach will give a fpras for reliability only in a restricted set of instances.

Now let f be a function from input strings to the natural numbers. A *randomised approximation scheme* for f is a probabilistic algorithm that takes as an input a string x and a rational number ϵ, $0 < \epsilon < 1$, and produces as output a random variable Y, such that Y approximates $f(x)$ within ratio $1 + \epsilon$ with probability $\geq 3/4$.

In other words,

$$(8.2.5) \qquad Pr\left\{\frac{1}{1+\epsilon} \leq \frac{Y}{f(x)} \leq 1+\epsilon\right\} \geq \frac{3}{4}.$$

It is important to note that the probability space here is not the set of possible inputs, but the 'random' computations.

Note also that the quantity 3/4 is not crucial and can be replaced by any number strictly greater than 1/2.

A *fully polynomial randomised approximation scheme* (fpras) for a function $f : \Sigma^* \to \mathbb{N}$ is a randomised approximation scheme which runs in time which is a polynomial function of n *and* ϵ^{-1}.

Suppose now we have such an approximation scheme and suppose further that it works in polynomial time. Then we can boost the success probability up to $1 - \delta$ for any desired $\delta > 0$, by using the following trick of Jerrum, Valiant and Vazirani (1986). This consists of running the algorithm $O(\log \delta^{-1})$ times and taking the median of the results.

We make this precise as follows:

(8.2.6) Proposition. If there exists a fpras for computing f then there exists an ϵ-δ approximation scheme for f which on input $\langle x, \epsilon, \delta \rangle$ runs in time which is bounded by $O(\log \delta^{-1})\mathrm{poly}(x, \epsilon^{-1})$.

Proof. Suppose $0 < \delta < 1$ and that $\mathcal{A}$ is a fpras for approximating f. Suppose that for a given x, ϵ, one runs t where $t = 12\lceil -\log \delta \rceil + 1$, independent simulations of the algorithm $\mathcal{A}$, yielding t random variables $Y_1, ..., Y_t$ each satisfying (8.2.5). Now if Y_{med} denotes the median of these t values the probability that Y_{med} fails to approximate $f(x)$ to within ratio $1 + \epsilon$ is bounded above by

$$\sum_{i=\frac{1}{2}(t+1)}^{t} \binom{t}{i} \left(\frac{1}{4}\right)^i \left(\frac{3}{4}\right)^{t-i}.$$

By Chernoff's inequality this bound is less than $e^{-t/12}$ and hence by choice of t, Y_{med} is an ϵ-δ-approximation to $f(x)$. $\qquad\square$

It is worth emphasising here that (8.2.5) shows that the existence of a fpras for a counting problem is a very strong result, it is the analogue of an *RP* algorithm for a decision problem and corresponds to the notion of tractability. However we should also note, that by an analogous argument to that used in proving Proposition (8.2.2) we have:

(8.2.7) Proposition: If $f : \Sigma^* \to \mathbb{N}$ is such that deciding if f is non-zero is NP-hard then there cannot exists a fpras for f unless NP is equal to random polynomial time RP.

Since this is thought to be most unlikely, it makes sense only to seek out a fpras when counting objects for which the decision problem is not NP-hard.

8.3 Generating solutions at random

For many problems the difficulty of generating a solution, uniformly at random, from the set of all solutions to a given instance appears to be roughly on a par with the difficulty of counting the solutions, or at least approximating the number of solutions. The fundamental paper on this topic is that of Jerrum, Valiant and Vazirani (1986) and it is on this that this section is based.

First an informal summary of what we will observe.

(8.3.1) Generating a random solution can be significantly harder than finding a solution.

(8.3.2) Generating a random solution can be significantly easier than counting all solutions.

To see that uniform generation may be easier than counting consider the following example.

(8.3.3) Example: Associate with any input consisting of a Boolean formula in disjunctive normal form (DNF) its set of satisfying assignments. We know from §1.4 that counting the number of satisfying assignments to a DNF is $\#P$-complete. However, there is a method of generating a random member of the solution set. It is based on the Karp-Luby (1983) method of approximating their number, and its running time is bounded by a polynomial. $\qquad \square$

Not surprisingly, it is usually harder to generate a random member of a population than to construct a single member of that same population.

It is easy to construct (= find) a directed cycle from a digraph whenever such a cycle exists. However we have:

(8.3.4) Proposition. If there exists a polynomial time bounded machine which constructs uniformly at random a directed cycle from a digraph then $NP = RP$.

Another theorem of Jerrum, Valiant and Vazirani shows that if it is possible to efficiently estimate the number of ways of extending a partial structure to a complete structure, then there is an efficient randomised algorithm for generating structures at random. Conversely, if there is an efficient algorithm for generating extensions of partial structures to complete structures then it is also possible to efficiently estimate the number of structures.

Finally, we mention the concept of approximate random generation.

We work with self reducible relations $R : \Sigma^* \times \Sigma^*$ The concept of self reducibility captures the idea that the solution set associated with a given instance of a problem can be expressed in terms of the solution sets of a number of smaller instances of the same problem.

Provided we are dealing with self reducible relations, a fast algorithm for approximate counting exists if and only if there exists a fast algorithm for almost uniform generation.

8.4   Rapidly mixing Markov chains

We return now to sampling procedures based on Markov chain simulation. Although the mathematical analysis can be very difficult the basic idea is straightforward.

Identify the set of objects being sampled with the state space Ω of a finite Markov chain constructed in such a way that the stationary distribution of the Markov chain agrees with the probability distribution on the set of objects. Now start the Markov chain at an arbitrary state and let it converge to the stationary distribution. Provided the convergence is fast, we have a good approximation to the stationary distribution fairly quickly. However the key to the success of this approach is that the time or number of steps needed to ensure that it is sufficiently close to its stationary distribution is not too large.

To be more specific, suppose that P is the transition matrix of an irreducible aperiodic finite Markov chain $(X_t; 0 \le t < \infty)$ with state space Ω. Suppose also that the chain is *reversible* in that it satisfies the detailed balance condition of (8.1.1).

This condition implies that π is a stationary or equilibrium distribution for P and that

$$Pr\{\lim_{t \to \infty} X_t = i\} = \pi_i$$

for all states $i \in \Omega$.

Moreover, P has eigenvalues $1 = \lambda_1 \ge \lambda_2 \ge \ldots \ge \lambda_N > -1$ where N is the number of states, and all these eigenvalues are real. The rate of convergence to the distribution π is determined by the quantity

$$\lambda_{\max} = \max\{\lambda_2, |\lambda_N|\}.$$

More precisely we have the following proposition. Let $P_{ij}(t)$ denote

$Pr\{X_{t+h} = j \mid X_h = i\}$ and let

$$\Delta_i(t) = \frac{1}{2}\sum_j |P_{ij}(t) - \pi(j)|,$$

denote what is called the *variation distance* at time t. It is clearly a measure of the separation from the stationary distribution at time t.

Now define, for $\epsilon > 0$, the function τ_i by

$$\tau_i(\epsilon) = \min\{t : \Delta_i(s) \le \epsilon \qquad \forall s \ge t\}.$$

(8.4.1) Proposition (Sinclair (1991)). For $\epsilon > 0$, $\tau_i(\epsilon)$ satisfies

(i) $\tau_i(\epsilon) \le \dfrac{1}{1 - \lambda_{max}}(ln\dfrac{1}{\pi(i)} + ln\dfrac{1}{\epsilon})$;

(ii) $\max_i \tau_i(\epsilon) \ge \dfrac{\lambda_{max}}{2(1 - \lambda_{max})} ln\left(\dfrac{1}{2\epsilon}\right)$.

In order to achieve rapid convergence we need $\tau_i(\epsilon)$ be small, for all i, and thus we need λ_{max} to be "relatively small".

Now λ_{max}, being dependent on two eigenvalues is a bit of a pain and it is useful to note the following trick which concentrates the interest on λ_2. Replace P by $P' = \frac{1}{2}(I + P)$ where I is the identity matrix. This only affects rates of convergence by a polynomial factor. All eigenvalues of P' are non-negative, and the quantity λ'_{max} of P' is $\frac{1}{2}\lambda_{max}$, so that henceforth we need only consider the second eigenvalue λ_2. Ideally we want $1 - \lambda_2$ to be large so λ_2 must be small.

The key idea of Sinclair and Jerrum (1989) was to relate this to the very aptly named *conductance* Φ of the chain. This is defined by

$$(8.4.2) \qquad \Phi = \max_{S \subseteq \Omega}\left\{\sum_{i \in S, j \in \Omega \backslash S} P(i,j)\pi(i) \bigg/ \sum_{i \in S} \pi(i)\right\}.$$

But the bracketed term is just the conditional probability of leaving a set S in the equilibrium state. In other words Φ is a measure of the ability of the chain to escape from any subset of the state space Ω.

(8.4.3) Theorem. The second eigenvalue λ_2 of a reversible ergodic Markov chain satisfies

$$1 - 2\Phi \le \lambda_2 \le 1 - \Phi^2/2.$$

In other words, for fast approximation we need large conductance.

Returning to Proposition 8.4.1 we see that combining it with Theorem 8.4.3 gives

$$\tau_i(\epsilon) \le \text{poly}(n),$$

where poly(n) denotes some polynomial function of the input size, if and only if

$$\Phi \geq \frac{1}{\text{poly}(n)}.$$

The first major practical application of the theory seems to have been in the work of Broder (1986) on approximating the permanent. The paper of Jerrum and Sinclair (1989) which built on Broder's work was a major advance, particularly in their idea of relating rapid mixing with conductance. They used this to show that there was a fpras for evaluating a dense permanent. More precisely they proved the following theorem.

We say that an $n \times n$ matrix A with entries from $\{0,1\}$ is *dense* if every row and column sum is at least $n/2$. Jerrum and Sinclair prove that

(8.4.4) Provided the matrix A is dense

$$\Phi \geq \frac{1}{12n^6}.$$

Using this in conjunction with (8.4.3) gives

(8.4.5) Theorem. There is a fpras for computing the permanent of dense matrices.

This still leaves the question open for matrices which are not dense, and resolving this is one of the more interesting open problems in this area. The techniques used to prove Theorem 8.4.5 are similar to those used to prove a similar result about the ferromagnetic Ising model. We will consider this in §8.6, but now turn to what is basically a more fundamental problem than the permanent.

8.5 Computing the volume of a convex body

The mathematical problems involved in computing areas and volume have a long history, they go back at least as far as 2000 B.C., (see for example Dyer and Frieze (1991)). Here we are concerned with a special case of the high dimensional integral discussed earlier, namely how to quickly compute the volume of a convex region of $\mathbf{R}^n$ where n is reasonably large.

Before we consider the specific computational question, we should emphasise that there can be severe difficulties in even describing the convex body. A good overview of the sort of problems which can arise is given in the article by Lovász (1990); a complete discussion may be found in the monograph of Grötschel, Lovász and Schrijver (1988). In the following very brief discussion of the volume question we shall restrict attention to convex bodies with algorithmically good descriptions.

First a negative result; we show that even when the region is a convex polyhedron with facets defined by hyperplanes whose coefficients are 0-1 the problem of computing the enclosed volume is $\#P$-hard.

Let $\prec$ be a partial order on the set $[n] = \{1, 2, ..., n\}$. Define the *order polytope* $P(\prec)$ to be the convex polyhedron in $\mathbf{R}^n$ consisting of the intersection of the unit hypercube C^n with

$$x : x_i \leq x_j \quad \text{if} \quad i \prec j.$$

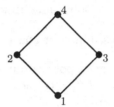

Figure 8.1.

Thus for example, if $\prec$ is the partial order shown in Figure 8.1 the associated polytope consists of $\mathbf{x} \in \mathbf{R}^4$ such that

$$0 \leq x_i \leq 1$$

$$x_1 \leq x_2 \leq x_4$$

$$x_1 \leq x_3 \leq x_4.$$

Now let $E(\prec)$ be the set of linear extensions of $\prec$.

(8.5.1) Proposition. Volume $(P(\prec)) = \#E(\prec)/n!$.

Proof. For any permutation π of $[1, n]$ let $S_\pi = \{x \in C^n : x_{\pi(1)} \leq x_{\pi(2)} \leq \dots x_{\pi(n)}\}$. Then use the following two, easily checked, results.

(8.5.2) The S_π are pairwise disjoint.

(8.5.3) The convex polytope $P(\prec) = \cup_{\pi \in E(\prec)} S_\pi$.

Also note that since the volume of a simplex Δ in $\mathbf{R}^n$ with vertices $p_0, p_1, ..., p_n$ is given by the determinantal formula

$$vol(\Delta) = \begin{vmatrix} 1 & 1 & \dots & 1 \\ p_0 & p_1 & \dots & p_n \end{vmatrix}$$

then we get that $vol(S_\pi) = 1/n!$ which proves (8.5.1). $\qquad\square$

In other words, computing the volume of the order polytope of a partial order $\prec$ reduces to enumerating the number of linear extensions of $\prec$.

Now we use the result of Brightwell and Winkler (1990) that counting the number of linear extensions is $\#P$-complete and we have the result:

(8.5.4) Theorem. Computing the volume of a convex polyhedron even when defined by (0-1) hyperplane constraints is $\#P$-hard.

At the same time, it is shown by Dyer and Frieze (1988) that the following is true.

(8.5.5) The volume of a convex polyhedron can be computed, to any polynomial number of bits, in polynomial time using a $\#P$-oracle.

An interesting question raised in Dyer and Frieze (1991) is the following.

(8.5.6) Problem. Suppose that a polyhedron is presented as a convex hull of a set of m points $p_1, ..., p_m \in \mathbf{R}^n$, with each p_i having 0-1 coordinates. Is computing the volume $\#P$-hard?

It is known that with general rational coordinates the problem is $\#P$-hard (see Dyer-Frieze (1988)), the above question is asking whether it is strongly $\#P$-hard.

In 1989, Dyer, Frieze and Kannan designed a fully polynomial time randomised algorithm to approximate the volume of a convex body $K \subseteq \mathbf{R}^n$. The original algorithm to find an ϵ-approximation to $vol(K)$ with $\epsilon < 1$ and error probability less than δ demanded

$$O(n^{23}(\log n)^5 \epsilon^{-2} \log(e^{-1}) \log(\delta^{-1}))$$

convex programmes.

Since then there have been improvements by Lovász and Simonovits (1990) and Applegate and Kannan (1991) which resulted in algorithms involving at most $O(n^{10})$ membership tests (though we are ignoring the additional terms in $(\log n)(\log \epsilon^{-1})$ and the like.

More recently, Dyer and Frieze (1991) have refined the algorithm so that currently computing the volume of $K \subseteq \mathbf{R}^n$ by a fpras involves $O(n^8 \epsilon^{-2} \log(n/\epsilon) \log \delta^{-1})$ membership tests.

Each of these successive improvements has a common theme, though the technical details are formidable. We attempt to give some insight into the nature of the proofs.

Proof Idea

Let K be a convex body in $\mathbf{R}^n$. Surround K by a ball $B \subseteq \mathbf{R}^n$ whose volume we know, and ideally with B, as close as possible to K. Then, even though the volume of K may be much smaller than B, it can be shown that there exists a sequence $K = K_0 \subseteq K_1 \subseteq ... \subseteq K_m = B$ of convex bodies where $vol(K_{i+1}) \leq 2vol(K_i)$ and m is only a polynomial in n. Thus provided each of the ratios $vol(K_i)/vol(K_{i+1})$ can be estimated, these estimates can be combined to give an estimate of $vol(K) = vol(K_0)$. In order to make this work we need to have a method of generating a point randomly with uniform distribution in a convex body. The rough idea of an algorithm for doing this is to construct a graph G, most of whose vertices are the lattice points of a grid or mesh of $\mathbf{R}^n$ which are contained in K. (Recall that a *lattice point* is a point in $\mathbf{R}^n$ having integral coordinates and a *cube* means a unit cube whose centre is a lattice point and whose edges are parallel to the axes.) Let V consist of those lattice points which are declared to be in K by a weak separation oracle using error tolerance $\frac{1}{2}$. The graph G has edges formed by joining two lattice points of V if and only if their distance apart is say d. G will have exponentially many nodes (about vol (K)) and maximum degree $2n$.

The idea now is to carry out a random walk on G and to use the conductance results of Sinclair and Jerrum described earlier to show that the associated Markov chain is rapid mixing.

To do this one needs good isoperimetric inequalities. However as Lovász (1990) remarks "the main problem is that the random walks are slow in getting to the distant parts of K". Thus at the moment, although in theory the problem has a fully polynomial randomised approximation, from a practical viewpoint the method needs to be speeded up.

However, the main and most important point is that the theoretical barrier has been broken, and one would expect substantial improvements in speed over the next few years.

8.6 Approximations and the Ising model

In §8.1 we have described a Metropolis type method for approximating the partition function of the random-cluster model. This includes the Ising model but the method does not converge fast.

We now describe another Metropolis algorithm which can be shown to consist of a rapidly mixing Markov chain and hence converges in polynomial time.

In order for the method to work we have to include a non-zero magnetic field, so that the Hamiltonian $H(\sigma)$ is given by

$$H(\sigma) = -J \sum_{ij \in E} \sigma_i \sigma_j - M \sum_k \sigma_k.$$

The idea is to construct a Markov chain whose states are spin configurations $\sigma \in \{-1, +1\}^n$ where $n = |V(G)|$. Transitions occur only between states which differ in one component. Transition probabilities are chosen so that the stationary distribution π is given by

(8.6.1) $\pi(\sigma) = \exp(-\beta H(\sigma))/Z.$

However, in order to have a fpras we have to work in a different scenario - what Jerrum and Sinclair (1990) call the "subgraphs-world". In this, the basic configurations are subsets of the edge set and the *weight* $w(X)$ of a given configuration X is calculated according to the formula:

:- a multiplicative factor λ for each edge of X
:- a multiplicative factor μ for each vertex of odd degree in X.
Then take a probability distribution

(8.6.2) $$Pr(X) = w(X)/Z' \quad X \subseteq E(G),$$

where Z' is the normalising factor. Taking

$$\lambda = \tanh \beta J$$
$$\mu = \tanh \beta M$$

gives

$$Z = 2^n (\cosh \beta M)^n (\cosh \beta J)^{|E|} Z'.$$

The method now is to set up a Markov chain whose states are configurations of this subgraphs world and whose stationary distribution is given by (8.6.1).

Transitions are allowed between states which differ in just one edge and transition probabilities from a state X to a new state X' are defined by the following model.

(8.6.3) Choose edge $e \in E$ uniformly at random and let $Y = X \oplus \{e\}$.

(8.6.4) If $w(Y) \geq w(X)$ then set $X' = Y$.

(8.6.5) If $w(Y) < w(X)$ then with probability $w(Y)/w(X)$ set $X' = Y$ otherwise $X' = X$.

The main result of Jerrum-Sinclair (1990) is the following:

(8.6.6) Theorem. The Markov chain on the subgraphs world is rapidly mixing.

There are a few points to note about the above method:

(a) we need a non-zero external field in order to make the Markov chain irreducible;

(b) we need $J > 0$, in other words ferromagnetism, otherwise the quantities in question are not probabilities;

(c) the calculations involved at each transition are not severe, to determine the ratio $w(Y)/w(X)$ we only need to carry out two multiplications since Y, X differ in just one edge.

We should note also that the relationship between the "subgraphs-world" and the more familiar "spins-world" of the Ising model goes back a long way in the physics literature, at least as far as Newell and Montroll (1953).

We now try to give some indication of the proof of Theorem 8.6.6.

Idea of Proof

As we have already indicated, the fundamental objective is to get good lower bounds on the conductance Φ of the Markov chain M defined by (8.6.3-5). To do this, the key step is to construct a *canonical path* γ_{xy} between each ordered pair of states x, y in the graph $G(M)$ (which represents the state space of the Markov chain). Provided these canonical paths can be chosen in such a way that no edge (transition) is overloaded by paths, then the chain cannot contain a bottleneck (or constriction). This implies that Φ cannot be too small, for if there were a bottleneck between S and $\Omega \backslash S$ then any choice of paths would overload the edges in the bottleneck. Thus we have shifted the problem from bounding eigenvalues, through conductances, to finding a "good" set of canonical paths having the property that the maximum loading of an edge $e = (u, v)$ of $G(M)$, measured by

$$(8.6.7) \qquad \rho = \max_e \frac{1}{Q(e)} \sum_{\gamma_{xy}} \pi(x)\pi(y)$$

is not too large.

Here

$$Q(e) = Q(u, v) = \pi(u)P(u, v)$$

and the sum in (8.6.7) is over all canonical paths from x to y which use e.

The relationship between ρ and conductance is that it can be shown that

$$\Phi \geq 1/2\rho.$$

The difficult part is to find a good set of canonical paths.

In the case of the Ising problem on $G = \langle V, E \rangle$, a collection of canonical paths is found for which

$$\rho \leq 2|E|/\mu^4.$$

This leads to a bound

$$\Phi \geq \frac{\mu^4}{4|E|}$$

and thus to

$$\lambda_2 \leq 1 - \frac{\mu^8}{32|E|^2}.$$

which is enough to show rapid mixing. $\qquad\qquad\qquad\qquad\qquad\qquad\qquad\qquad\square$

8.7 Some open questions

A natural question to ask, in view of the Jerrum-Sinclair fpras for the ferromagnetic Ising problem is whether there exists a fpras for the Q-state Potts model, again restricted to the ferromagnetic case. Specifically we pose

(8.7.1) Problem. Is there a fpras for the Q-state Potts model in the ferromagnetic case for $Q \geq 3$?

One might ask the more specific question whether the algorithm of Swendsen-Wang (1987) is underpinned by a rapidly mixing Markov chain. This seems hard to prove, even if it is true. However there is such a structural similarity between the Q-state Potts for general integer Q and the Ising special case ($Q = 2$) that an affirmative answer to (8.7.1) seems possible.

In the same way as Jerrum and Sinclair (1990) show that there is no fpras for the antiferromagnetic Ising model unless $NP = RP$, it can be shown that the following is true.

(8.7.2) There is no fpras for the antiferromagnetic Potts model for $Q \geq 3$, unless $NP = RP$.

However, I have been unable to resolve the following related question.

(8.7.3) Problem. Does there exist a fpras for counting the number of 4-colourings of a planar graph?

A similar question about 3-colourings is easily answered. It is known to be NP-hard to decide if a planar graph has a 3-colouring. Hence there cannot exist a fpras for counting 3-colourings in a planar graph unless $NP = RP$.

A very interesting general question posed in Broder (1986) and which is related to some of the questions posed above is the following.

(8.7.4) Problem. Does there exist a problem for which a solution can be found in polynomial time, but for which approximating the number of solutions is hard.

I believe that an answer to this may be found in 4-colouring planar graphs. Finding a 4-colouring is in P by the method of proof of the Four Colour Theorem. However I believe that it is NP-hard to approximately count the 4-colourings. In other words, I believe that the answer to (8.7.3) is negative and go further by making, perhaps rashly, the following:

(8.7.5) Conjecture. For any fixed $k \geq 4$ there is no fpras for counting the number of k-colourings of a planar graph.

Conjecture (8.7.5) is expressible in terms of flows rather than colourings and in this connection the following result of Mihail and Winkler (1991) is very interesting.

Let every vertex of G have even degree. Then an *Eulerian orientation* of G is an assignment of directions to the edges so that for each vertex the number of edges directed inward equals the number of edges directed outward. Let $e(G)$ denote the number of such orientations. This is clearly closely related to the ice problem discussed in §3.6, and hence to the general flow problem. The main results of Mihail and Winkler are the following.

(8.7.6) Computing $e(G)$ is #P-complete.

(8.7.7) There exists a fpras for determining $e(G)$.

Now when G is regular of degree 4, it is clear that $e(G)$ is just the number of ice configurations on G, in other words it is essentially the number of 3-flows of G. Thus a special case of (8.7.6) and one which is of particular interest to us is the following.

(8.7.8) For the class of regular graphs of degree 4 there is a fpras for computing the number of ice configurations.

For planar graphs the statement (8.7.7) can be dualised to give the following result.

(8.7.9) There exists a fpras for determining the number of 3-colourings of a planar graph in which each face has 4 edges.

This is somewhat surprising, and suggests that it may be possible to extend the arguments to answer positively the following question.

(8.7.10) Does there exist a fpras for counting 4-colourings of planar graphs of bounded vertex degree?

The arguments of Mihail and Winkler are based on a transformation from the matching arguments used to prove a fpras for dense permanent and do not seem easy to modify to deal with (8.7.10).

Turning now to a more practical question; this is whether the reliability probability can be approximated.

Recall that this is essentially a realisation of the Tutte polynomial along the degenerate hyperbola $x = 1$ and it would seem to be difficult to disprove the following conjecture.

(8.7.11) Conjecture. There exists a fpras for computing the reliability probability of a planar graph.

Indeed, there seems no reason to restrict the above conjecture to the planar case except that it was much easier to prove hardness in the non-planar case and so conversely it may be easier to deal with the planar case first.

Dualising, a special case of (8.7.11) and one which seems to contain many of the ingredients which make it a key problem in this area is the following.

(8.7.12) Problem. Does there exist a fpras for counting the number of forests in a (planar) graph?

A negative answer to this would certainly settle (8.7.11), a positive answer to (8.7.11) would probably lead to a positive answer to the reliability question.

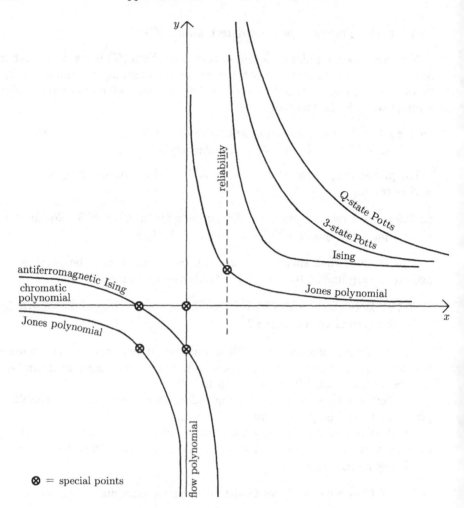

Figure 8.2. Map of Tutte plane and its specialisations.

A general conjecture which, if true, encompasses most of what has gone before, is the following. Consider the Tutte plane as illustrated in Figure 8.2

(8.7.13) Conjecture. There exists a fpras for computing the evaluation of the Tutte polynomial of a graph at each (algebraic) point (x, y) of the positive quadrant $x \geq 1$, $y \geq 1$.

For technical reasons, based on a study of the random cluster model and on a certain "well behaved log concavity" (see Seymour and Welsh (1975)), I have greater faith in the truth of (8.7.13) for the region $x \geq 1$, $y \geq 1$. However there is no clear obstacle to it being true in the whole positive quadrant. A borderline problem is the following which amounts to an evaluation of $T(G; 2, 0)$.

(8.7.14) Problem. Does there exist a fpras for counting the number of acyclic orientations of a (planar) graph?

Finally note that in many of the above cases I have highlighted planarity as a possible separator between easy and hard cases.

The Robertson-Seymour theory of minors leads to the following conclusions. Almost all problems are "easy", that is in P, for graphs of bounded tree-width. The square lattice or "grid graph" is the minimal graph with unbounded tree width. This suggests the possibility that when the underlying graph is of lattice type the computational problems are only just the wrong side of the feasibility barrier.

Accordingly we are led to the following fundamental question.

(8.7.16) Problem. Let $\mathcal{L}_n$ be the $n \times n$ toroidal square lattice. Is there a constant c such that computing or finding a fpras for the Tutte polynomial $T(\mathcal{L}_n; x, y)$ for integer (x, y) can be carried out in time $O(n^c)$?

Settling this problem would be a major advance both theoretically and practically.

8.8 Additional notes

The monograph of Hammersley and Handscombe (1964) gives a wide ranging authoritative account of Monte Carlo methods.

The discussion on Monte-Carlo methods in §8.1 is based on Sokal (1989). Example 8.1.3 is my attempt to simulate the random cluster process but its convergence seems to be slow. The streamlined version of the stochastic Potts described in Example 8.1.2 is very close to the model underlying the colouring algorithm of Petford and Welsh (1989). Even though this is a finite problem it displays "a critical phenomenon" but as a practical colouring method seems to be very effective. For a discussion of reversibility see Kelly (1979). Aldous (1982) and Aldous and Diaconis (1986) are good sources for the sort of questions discussed in §8.4. Proposition (8.4.1) is an enhanced version of an earlier result of Sinclair and Jerrum (1989). The #P- hardness of volume computation was first proved by Bárány and Füredi (1986). Lovász (1990) gives an elegant account of the methods and difficulties involved in computing volumes in his general survey of geometric algorithms. This article and that of Dyer and Frieze (1991) are the basis of most of the material in §8.5.

A Monte-Carlo approach to reliability is taken by Karp and Luby (1985); this approach however depends on the failure probabilities being sufficiently small. Karpinski and Luby (1989) develop a fpras for the number of zeros of a multivariate polynomial over $GF(2)$; the exact counting problem is #P-hard even when restricted to cubic polynomials. This problem, is basically a parity version of the DNF counting problem – the object is to count the number of assignments which satisfy an odd number of terms.

Returning finally to a favourite problem, namely that of enumerating self-avoiding walks on the square lattice. This is becoming a benchmark of computer performance and algorithm design. Conway, Enting and Guttmann (1992) have now counted walks up to 39 steps, as against 24, which was the state of the art in 1972. Using this, Conway and Guttmann (1992) have proved that $e^\kappa \geq 2.62$. This is less than 0.7% below the best numerical estimate of 2.63815853, and illustrates again, that even if the general case may be $\#P$-hard, good asymptotic results can be obtained.

References

Adyan, S.I. (1958) On algorithmic problems in effectively complete classes of groups, *Dokl. Akad. Nauk. SSSR*, **123**, 13-16.

Aigner, M. (1979) *Combinatorial Theory*, Springer-Verlag, Berlin.

Aizenman, M., Chayes, J.T., Chayes, L., and Newman, C.M. (1988) Discontinuity of the magnetization in one-dimensional $1/|x-y|^2$ Ising and Potts models. *J. Stat. Phys.* **50**, 1-40.

Akutsu, Y. and Wadati, M. (1987) Exactly solvable models and new link polynomials. I. N-state vertex models, *J. Phys. Soc. Jpn.* **56**, 3039-3051.

Aldous, D. (1982) Some inequalities for reversible Markov chains, *J. Lond. Math. Soc. (2)* **25**, 564-576.

Aldous, D. and Diaconis, P. (1986) Shuffling cards and stopping times, *Amer. Math. Monthly* **93**, 333-348.

Alexander, J.W. (1923) A lemma on systems of knotted curves, *Proc. Nat. Acad. Sci. USA* **9**, 93-95.

Alexander, J.W. (1928) Topological invariants of knots and links, *Trans. Amer. Math. Soc.* **30**, 275-306.

Allender, E. (1985) *Invertible functions.* PhD thesis, Georgia Tech.

Anger, A.L. (1959) *Machine calculation of knot polynomials*, Princeton Senior Thesis.

Annan, J. (1993) (personal communication).

Anstee, R.P., Przytycki, J.H., Rolfsen, D. (1989) Knot polynomials and generalized mutation, *Topology and Applications* **32**, 237-249.

Applegate, D. and Kannan, R. (1991) Sampling and integration of near log-concave functions, *Proc. 23rd ACM Symp. Th. Comput.*, 156-163.

Ashkin, J. and Teller, E. (1943) Statistics of two-dimensional lattices with four components, *Phys. Rev.* **64**, 178-184.

Atiyah, M.F. (1990) *The Geometry and Physics of Knots*, Cambridge University Press.

Babai, L. and Fortnow, L. (1990) A characterization of $\#P$ by arithmetic straight line programs, *Proc. 31st IEEE Symp. Found. Comp. Sc.* 26-36.

Ball, M.O. (1977) The complexity of network reliability computations, *Networks* **10**, 153-165.

Ball, M.O. and Provan, J.S. (1983) Calculating bounds on reachability and connectedness in stochastic networks, *Networks* **13**, 253-278.

Balcázar, J.L., Diaz, J. and Gabarró, J. (1988) *Structural Complexity I* Springer-Verlag, Berlin.

Barahona, F. (1982) On the computational complexity of Ising spin glass models, *J. Phys. A* **15**, 3241-3253.

Bárány, I. and Füredi, Z. (1986) Computing the volume is difficult, *Proc. 18th Annual ACM Symposium on Theory of Computing*, 442-447

Baxter, R.J. (1982) *Exactly Solved Models in Statistical Mechanics*, Academic Press, London.

Beigel, R. (1990) Relativizing counting classes: Relations among thresholds, parity, and mods, *J. Comput. System Sci.* **42**, 76-96.

Beigel, R., Gill, J. and Hertrampf, U. (1990) Counting classes: Thresholds, parity, mods and fewness, *Proc. 7th Symp. on Theoretical Aspects of Computer Science*, Lecture Notes in Comput. Sci. 415, Springer-Verlag, New York, 49-57.

Berman, L. and Hartmanis, J. (1977) On isomorphisms and density of NP and other complete sets, *SIAM J. Comput.* **6**, 305-322.

Bezuidenhout, C.F., Grimmett, G.R., Kesten, H. (1992) Strict inequality for critical values of Potts models and random-cluster processes, (preprint).

Biggs, N. (1974) *Algebraic Graph Theory*, Cambridge Univ. Press.

Biggs, N. (1977) *Interaction Models*, London Math. Soc. Lecture Notes, Cambridge University Press.

Birkhoff, G.D. (1912) A determinant formula for the number of ways of coloring a map, *Ann. Math.* (2) **14**, 42-46.

Birkhoff, G.D. and Lewis, D.C. (1946) Chromatic polynomials, *Trans. Amer. Math. Soc.* **60**, 355-451.

Birman, J.S. (1974) Braids links and mapping class groups, *Ann. Math. Studies* 82, Princeton N.J., Princeton Univ. Press.

Blass, A. and Gurevich, Y. (1982) On the unique satisfiability problem, *Inform. and Control.* **55**, 80-88.

Bollobás, B. (1979) *Graph Theory*, Springer-Verlag, Berlin.

Brandt, R.D., Lickorish, W.B.R. and Millett, K.C. (1986) A polynomial invariant for unoriented knots and links, *Invent. Math.* **84**, 563-573.

Brightwell, G. and Winkler, P. (1991) Counting linear extensions is $\#P$-complete, *Proc 23rd Symp. Th. Comp.*, 175-182.

Broadbent, S.R. and Hammersley, J.M. (1957) Percolation processes I. Crystals and mazes, *Proc. Camb. Phil. Soc.* **53**, 629-641.

Broder, A.Z. (1986) How hard is it to marry at random? (On the approximation of the permanent), *Proceedings of the 18th ACM Symposium on Theory of Computing*, 1986, pp.50-88. Erratum in *Proceedings of the 20th ACM Symposium on Theory of Computing*, 1988, p.551.

Brylawski, T.H. (1972) A decomposition for combinatorial geometries, *Trans. Amer. Math. Soc.* **171**, 235-282.

Brylawski, T.H. (1980) The Tutte polynomial, matroid theory and its applications. Centro Internazionale Matematico Evisto 3 Liguori, Naples, 125-275.

Brylawski, T.H. and Oxley, J.G. (1992) The Tutte polynomial and its applications, *Matroid Applications* (ed. N. White), Cambridge Univ. Press, 123-225.

Burde, G. and Zieschang, H. (1985) *Knots*, de Gruyter.

Cai Jin-yi and Hemachandra, L.A. (1989) On the power of parity polynomial time, *Proc. Symposium on Theoretical Aspects of Computer Science*, Lecture Notes in Computer Science **349**, Springer-Verlag, 229-240.

Cipra, B.A. (1987) An introduction to the Ising model, *American Math. Monthly* 94, 937-959.

Colbourne, C.J. (1987) *The Combinatorics of Network Reliability*, Oxford University Press.

Conway, J.H. (1969) An enumeration of knots and links, *Computational problems in abstract algebra* (ed. J. Leech), Pergamon Press, 329-358.

Conway, A.R., Enting, I.G. and Guttmann, A.J. (1992) Algebraic techniques for enumerating self-avoiding walks on the square lattice, *J. Phys. A: Math. Gen.* (to appear)

Conway, A.R. and Guttmann, A.J. (1992) Lower bound on the connective constant for square lattice self avoiding walks (preprint).

Cook, S.A. (1971) The complexity of theorem-proving procedures, *Proc. 3rd Ann. ACM Symp. on Theory of Computing*, 151-158.

Crapo, H. (1969) The Tutte polynomial, *Aequationes Math.* 3, 211-229.

Crowell, R.H. and Fox, R.H. (1977) *Introduction to Knot Theory*, Springer-Verlag, Berlin.

Dietrich-Buchecker, C.D. and Sauvage, J.P. (1989) A synthetic molecular trefoil knot, *Angew, Chem. Int. Ed.* 28, 189-192.

Dyer, M.E. and Frieze, A. (1988) On the complexity of computing the volume of a polyhedron, *SIAM J. Comput.* 17, 967-974.

Dyer, M. and Frieze, A. (1991) Computing the volume of convex bodies: a case where randomness provably helps (preprint: January 1991).

Dyer, M., Frieze, A. and Kannan, R. (1989) A random polynomial time algorithm for approximating the volume of convex bodies. *Proceedings of the 21st ACM Symposium on Theory of Computing*, 375-381.

Edmonds, J., Johnson, E.L. (1973) Matching, Euler tours and the Chinese postman, *Math. Prog.* 5, 88-124.

Edwards, K.J. and Welsh, D.J.A. (1983) On the complexity of uniqueness problems (unpublished).

Edwards, R.G. and Sokal, A.D. (1988) Generalization of the Fortuin-Kasteleyn-Swendesen-Wang representation and Monte Carlo algorithms. *Phys. Rev. D* 38, 2009-2012.

Erdös, P. and Spencer, J. (1974) *Probabilistic Methods in Combinatorics*, Academic Press, New York.

Ernst, C. and Sumners, D.W. (1987) The growth of the number of prime knots, *Math. Proc. Camb. Phil. Soc.* 102, 303-315.

Ernst, C. and Sumners, D.W. (1990) A calculus for rational tangles: applications to DNA recombination. *Math. Proc. Camb. Phil. Soc.* 108, 489-.

Essam, J.W. (1971) Graph theory and statistical physics, *Discrete Math.* 1, 83-112.

Fisher, M.E. (1961a) Statistical mechanics of dimers on a plane lattice, *Phys. Rev.* 124, 1664-1672.

Fisher, M.E. (1961) Critical probabilities for cluster size and percolation problems, *J. Math. Phys.* 2, 620-627.

Fisher, M.E. (1966) On the dimer solution of planar Ising models, *J. Math. Phys.* 7, 1776-1781.

Fortuin, C.M. (1972) On the random cluster model. II. The percolation model, *Physica* 58, 393-418.

Fortuin, C.M. and Kasteleyn, P.W. (1972) On the random cluster model. I. Introduction and relation to other models, *Physica* 57, 536-564.

Fortuin, C.M., Kasteleyn, P.W., and Ginibre, J. (1971) Correlation inequalities on some partially ordered sets. *Comm. Math. Phys.* 22, 89-103.

Fox, R.H. (1962) A quick trip through knot theory, *Topology of 3- Manifolds* (ed. M.K. Fort, Jr.) Prentice-Hall, Englewood Cliffs, 120-167.

Fraenkel, A.S. and Loebl, M. (1992) Complexity of circuit intersections in graphs (to appear).

Frank, A. (1989) Conservative weightings and ear-decompositions of graphs, *Combinatorica* (to appear).

Freyd, P., Yetter, D., Hoste, J., Lickorish, W.B., Millett, K. and Ocneanu, A. (1985) A new polynomial invariant of knots and links, *Bull. Amer. Math. Soc.* 12, 239-246.

Frisch, H.L. and Wasserman, E. (1968) Chemical topology, *J. Amer. Chem. Soc.* 83, 3789-3795.

Gandolfi, A., Keane, M., and Newman, C.M. (1992) Uniqueness of the infinite component in a random graph with applications to percolation and spin glasses, *Probability Theory and Related Fields* 92, 511-527.

Garey, M.R. and Johnson, D.S. (1979) *Computers and Intractability — A guide to the theory of NP-completeness.* (San Francisco, Freeman)

Garside, F.A. (1969) The braid group and other groups, *Quart. J. Math. Oxford* (2) 20, 235-254.

Gauss, C.F. (1833) Zur mathematischen Theorie der electrodynamischen Wirkungen, *Werke Königl. Gesell. der Wissenchaften zu Göttingen* 5, 605.

Gill, J. (1977) Computational complexity of probabilistic Turing machines, *SIAM J. Computing* 6, 675-695.

Goldschlager, L.M. and Parberry, I. (1986) On the construction of parallel computers from various bases of Boolean functions. *Theor. Comput. Sci.* 43, 43-58.

Gordon, C. Mc A. and Luecke, J. (1989) Knots are determined by their complements, *Bull. Amer. Math. Soc.* 20, 83-87.

Greene, C. (1976) Weight enumeration and the geometry of linear codes, *Studies Appl. Math.* 55, 119-128.

Grimmett, G.R. (1989) *Percolation.* Springer-Verlag, Berlin.

Grollmann, S. and Selman, A.L. (1984) Complexity measures for public-key crypto-systems, *25th IEEE Symp. Found. Comput. Sci.*, 495-503.

Grötschel, M., Lovász, L. and Schrijver, A. (1988) *Geometric Algorithms and Combinatorial Optimization.* (Springer-Verlag, Berlin.)

Haken W. (1961) Theorie der Normalflächen, *Acta Math.* 105, 245-375.

Hammersley, J.M. (1957) Percolation processes II. The connective constant, *Proc. Cambridge Philos. Soc.* 53, 642-645.

Hammersley, J.M. (1961) Comparison of atom and bond percolation, *J. Math. Phys.* 2, 728-733.

Hammersley, J.M. (1991) Self avoiding walks, *Physica* A 177, 51-57.

Hammersley, J.M. and Handscombe, D.C. (1964) *Monte Carlo Methods*, Methuen, London.

Hammersley, J.M. and Mazzarino, G. (1983) Markov fields, correlated percolation and the Ising model, *The Mathematics and Physics of Disordered Media*; Springer Lecture Notes in Mathematics 1035, 201-245.

Hammersley, J.M. and Welsh, D.J.A. (1962) Further results on the rate of convergence to the connective constant of the hypercubical lattice, *Quart. J. Math. Oxford Ser. (2)* 13, 108-110.

Hansen, V.G. (1989) *Braids and Covering: Selected Topics*, Cambridge Univ. Press, Cambridge.

Hara, T. and Slade, G. (1991) Critical behaviour of self-avoiding walk in five or more dimensions, *Bull. Amer. Math. Soc.* **25**, 417-423.

de la Harpe, P., Kervaire, M. and Weber, C. (1986) On the Jones polynomial, *Enseign. Math. (2)* **32**, 271-335.

Hemachandra, L.A. (1987) *Counting in Structural Complexity Theory.* Ph.D. thesis, Cornell University.

Hemion, G. (1979) On the classification of homeomorphisms of 2-manifolds and the classification of 3-manifolds. *Acta Math.* **142**, 123-155.

Hemion, G. (1992) *The Classification of Knots and 3-Dimensional Spaces*, Oxford Univ. Press.

Hoggar, S. (1974) Chromatic polynomials and logarithmic concavity, *J. Comb. Theory (B)* **16**, 248-254.

Holyer, I. (1981) The *NP*-completeness of edge colouring, *SIAM J. Comput.* **10**, 718-720.

Hoste, J. (1986) A polynomial invariant for knots and links, *Pacific J. Math.* **124**, 295-320.

Hintermann, A., Kunz, H. and Wu, F.Y. (1978) Exact results for the Potts model in two dimensions, *J. Stat. Phys.* **19**, 623-632.

Holley, R. (1974) Remarks on the FKG inequalities. *Comm. Math. Phys.* **36**, 227-231.

Jaeger, F. (1976) On nowhere-zero flows in multigraphs, *Proc. Fifth. British Combinatorial Conference* (ed. C.St.J.A. Nash-Williams and J. Sheehan) Utilitas Math. Winnipeg, 373-378.

Jaeger, F. (1988a) Nowhere zero flow problems. *Selected Topics in Graph Theory* 3, (ed. L. Beineke and R.J. Wilson) Academic Press 71-92.

Jaeger, F. (1988b) On Tutte polynomials and link polynomials. *Proc. Amer. Math. Soc.* **103**, 647-654.

Jaeger, F. (1992) On the Kauffman polynomial of planar matroids, *Proc. Fourth Czechoslovak Symposium in Combinatorics*, Prachatice, to appear.

Jaeger, F., Vertigan, D.L., Welsh, D.J.A. (1990) On the computational complexity of the Jones and Tutte polynomials, *Math. Proc. Camb. Phil. Soc.* **108**, 35-53.

Janse van Rensburg, E.J. and Whittington, S.G. (1990) The knot probability in lattice polygons, *J. Phys. A. Math. Gen.* **23**, 3573-3590.

Jerrum, M.R. (1981) On the complexity of evaluating multivariate polynomials, Ph.D. thesis, Edinburgh.

Jerrum, M.R. (1987) 2-dimensional monomer-dimer systems are computationally intractable, *J. Stat. Phys.* **48**, 121-134.

Jerrum, M. and Sinclair, A. (1989) Approximating the permanent, *SIAM J. Comp.* **18**, 1149-1178.

Jerrum, M.R. and Sinclair, A. (1990) Polynomial-time approximation algorithms for the Ising model, *Proc. 17th ICALP, EATCS*, 462-475.

Jerrum, M.R., Valiant, L.G. and Vazirani, V.V. (1986) Random generation of combinatorial structures from a uniform distribution, *Theor. Comp. Sci.* **43**, 169-188.

Johnson, D.S. (1985) The *NP*-completeness column: an ongoing guide (15th edition), *J. Algorithms* **6**, 291-305.

Johnson, D.S. (1990) A catalog of complexity classes, *Handbook of Theoretical Computer Science*, Chapter 2, 68-161 (ed. J. van Leeuwen) Elsevier.

Jones, V.F.R. (1985) A polynomial invariant for knots via Von Neumann algebras, *Bull. Amer. Math. Soc.* 12, 103-111.

Jones, V.F.R. (1987) Hecke algebra representations of braid groups and link polynomials, *Ann. Math.* **126**, 103-112.

Jones, V.F.R. (1989) On knot invariants related to some statistical mechanical models, *Pacific J. Math.* **137**, 311-334.

Jones, V.F.R. (1991) *Subfactors and Knots*, Amer. Math. Soc. Providence, Rhode Island.

Karp, R.M. (1972) Reducibility among combinatorial problems, *Complexity of Computer Computations*, eds: R.E. Miller and J.W. Thatcher, Plenum Press, New York 85-103.

Karp, R.M. and Luby, M. (1983) Monte-Carlo algorithms for enumeration and reliability problems, *Proceedings 24th IEEE Symp. Found. Comp. Sci.*, pp. 56-64.

Karp, R.M. and Luby, M. (1985) Monte-Carlo algorithms for the planar multiterminal network reliability problem, *J. Complexity* **1**, 45-64.

Karp, R.M., Luby, M. and Madras, N. (1989) Monte-Carlo approximation algorithms for enumeration problems, *J. Algorithms* **10**, 429-448.

Karpinski, M. and Luby, M. (1989) Approximating the number of solutions of a $GF(2)$ polynomial (preprint).

Kasteleyn, P.W. (1961) The statistics of dimers on a lattice, *Physica* **27**, 1209-1225.

Kasteleyn, P.W. (1963) Dimer statistics and phase transitions, *J. Math. Phys*, **4**, 287-293.

Kasteleyn, P.W. (1967) Graph theory and crystal physics. *Graph theory and Theoretical Physics*, ed. F. Harary (London, Academic Press), 43-110.

Kauffman, L.H. (1987a) State models and the Jones polynomial, *Topology* **26**, 395-407.

Kauffman, L.H. (1987b) *On Knots*, Princeton Univ. Press.

Kauffman, L.H. (1988) New invariants in the theory of knots, *Am. Math. Monthly* **95**, 195-242.

Kauffman, L.H. (1989) A Tutte polynomial for signed graphs, *Discrete Applied Math.* **25**, 105-127.

Kauffman, L.H. (1990) An invariant of regular isotopy, *Trans. Amer. Math. Soc.* **318**, 417-471.

Kelly, F.P. (1979) *Reversibility and Stochastic Networks*, Wiley, New York.

Kesten, H. (1963) On the number of self-avoiding walks, *J. Math. Phys.* **4**, 960-969.

Kesten, H. (1980) The critical probability of bond percolation on the square lattice equals $\frac{1}{2}$, *Comm. Math. Phys.* **74**, 41-59

Kesten, H. (1982) *Percolation Theory for Mathematicians*, Birkhauser, Boston.

Kidwell, M.W. (1987) On the degree of the Brandt-Lickorish-Millett-Ho polynomial of a link, *Proc. Amer. Math. Soc.* **100**, 755-762.

Kilpatrick, P.A. (1975) *Tutte's first colour-cycle conjecture*, PhD thesis, Cape Town.

Köbler, J. Schöning and Toran, J. (1989) On counting and approximation, *Acta Informatica* **26**, 363-379.

Laanit, L., Messager, A., Miracle-Sole, S., Ruiz, J. and Shlosman, S. (1991) Interfaces in the Potts model 1: Pirogov-Sinai theory of the Fortuin-Kasteleyn representation. *Comm. Math. Phys.* **140**, 81-92.

Leven, D. and Galil, Z. (1983) NP-completeness of finding the chromatic index of regular graphs, *J. Algorithms* **4**, 35-44.

Lickorish, W.B.R. (1988) Polynomials for links, *Bull. Lond. Math. Soc.* **20**, 558–588.

Lickorish, W.B.R. and Millett, K.C. (1987) A polynomial invariant of oriented links, *Topology* **26**, 107-141.

Lickorish, W.B.R. and Millett, K.C. (1988) The new polynomial invariants of knots and links, *Math. Magazine* **61**, 3-23.

Lickorish, W.B.R. and Thistlethwaite, M.B. (1988) Some links with non-trivial polynomials and their crossing numbers, *Comment. Math. Helvetici* **63**, 527-539.

Lieb, E.H. (1967) Residual entropy of square ice, *Phys. Rev.* **162**, 162-171.

Linial, N. (1986) Hard enumeration problems in geometry and combinatorics, *SIAM J. Alg. Disc. Math.* **7**, 331-335.

Lovász, L. (1979a) *Combinatorial Problems and Exercises*, North Holland, Amsterdam.

Lovász, L. (1979b) On determinants, matchings and random algorithms. *Fundamentals of Computing Theory* (ed. L. Budach) Akademia-Verlag, Berlin.

Lovász, L. (1990) Geometric algorithms and algorithmic geometry, *Proc. Int. Cong. Math., (Kyoto) Japan*, 139-154.

Lovász, L. and Plummer, M.D. (1986) *Matching Theory*, Akademiai Kiado, Budapest.

Lovász, L. and Simonovits, M. (1990) The mixing rate of Markov chains, an isoperimetric inequality, and computing the volume; *Proc. 31st IEEE Symp. Found. Comp. Sc.*, 346-355.

Lyndon, R.C. and Schupp, P.E. (1977) *Combinatorial Group Theory*, Springer-Verlag.

Makanin, G.S. (1968) The conjugacy problem in the braid groups, *Soviet Math. Doklady* **9**, 1156-1157.

Mathon, R. (1979) A note on the graph isomorphism counting problem, *Inform. Proc. Lett.* **8**, 131-132.

Menasco, W.W. and Thistlethwaite, M.B. (1991) The Tait flyping conjecture. *Bull. Amer. Math. Soc.* **25**, 403-412.

Mihail, M. and Winkler, P. (1991) On the number of Eulerian orientations of a graph, Bellcore Technical Memorandum TM-ARH-018829.

Moore, E.F. and Shannon, C.E. (1956) Reliable circuits using less reliable components, *Journ. Franklin Instit.* **262**, 191-208 and 281-297.

Morton, H.R. and Short, H.B. (1990) Calculating the 2-variable polynomial for knots presented as closed braids, *J. Algorithms* 11(1), 117-131.

Murasugi, K. (1987) Jones polynomials and classical conjectures in knot theory, *Topology* **26**, 187–194.

Murasugi, K. (1989) On invariants of graphs with applications to knot theory, *Trans. Amer. Math. Soc.* **314**, 1 49.

Murasugi, K. and Przytycki, J.H. (1991) The index of a graph with applications to knot theory (preprint).

Newell, G.F. and Montroll, E.W. (1953) On the theory of the Ising model of ferromagnetism, *Rev. Mod. Phys.* **25**, 353-389.

Oxley, J.G. (1992) *Matroid Theory*, Oxford Univ. Press.

Oxley, J.G. and Welsh, D.J.A. (1979) The Tutte polynomial and percolation, *Graph Theory and Related Topics* (eds. J.A. Bondy and U.S.R. Murty), Academic Press, London, 329-339.

Oxley, J.G. and Welsh, D.J.A. (1992) Tutte polynomials computable in polynomial time, *Discrete Math.* **109**, 185-192.

Papadimitriou and Yannakakis, M. (1984) The complexity of facets (and some facets of complexity), *J. Comput. System Sci.* **28**, 244-259.

Papadimitriou, C.H. and Zachos, S. (1983) Two remarks on the power of counting, *Proc. 6th GI conference on TCS*, Lecture Notes in Computer Science **145**, Springer-Verlag, 296-276.

Paterson, M.S. and Razborov, A.A. (1991) The set of minimal braids is *CO-NP*-complete, *J. Algorithms* **12**, 393-408.

Percus, J.K. (1971) *Combinatorial Methods* Springer-Verlag, New York.

Petford, A.D. and Welsh, D.J.A. (1989) A randomised 3-colouring algorithm, *Discrete Math.* **74**, 253-261.

Pippenger, N. (1989) Knots in random walks, *Discrete Appl. Math.* **25**, 273-278.

Potts, R.B. (1952) Some generalized order-disorder transformations, *Proc. Cambridge Philos. Soc.* **48**, 106-109.

Pratt, V. (1975) Every prime has a succinct certificate, *SIAM J. Comput.* **4**, 214-220.

Preston, C.J. (1974) *Gibbs States on Countable Graphs.* Cambridge University Press, Cambridge.

Provan, J.S. and Ball, M.O. (1983) On the complexity of counting cuts and of computing the probability that a graph is connected, *SIAM J. Computing* **12**, 777-788.

Provan, J.S. (1986) The complexity of reliability computations in planar and acyclic graphs, *SIAM J. Computing* **15**, 694-702.

Przytycka, T.M. and Przytycki, J.H. (1992) Subexponentially computable truncations of Jones-type polynomials with appendix on Vertigan's algorithm (preprint October 1992).

Rabin, M.O. (1958) Recursive unsolvability of group theoretic problems, *Ann. Math.* **67**, 172-174.

Read, R.C. (1968) An introduction to chromatic polynomials, *J. Combinatorial Theory* **4**, 52-71.

Reidemeister, K. (1935) Homotopieringe und Linsemräume, *Abh. Math. Sem. Hamburg* II, 102-109.

Reidemeister, K. (1948) *Knotentheorie* (Chelsea, New York); English translation, edited by L.F. Boron, C.D. Christenson and B.A. Smith, *Knot Theory* BCS Associates, Moscow, Idaho, 1983.

Robertson, N. and Seymour, P.D. (1985-present) Graph Minors I-XVII, *J. Combinatorial Theory* (Ser B).

Rolfsen, D. (1976) *Knots and Links*, Publish or Perish, Berkeley.

Rosenstiehl, P. and Read, R.C. (1978) On the principal edge tripartition of a graph, *Ann. Discrete Math.* **3**, 195-226.

Schöning, U. (1990) The power of counting, *Complexity Theory Retrospective* (ed. A.L. Selman), Springer-Verlag, Berlin, 204-222.

Schubert, H. (1949) Die endeutige Zerlegbarkeit eines Knoten in Primknoten, *Sitzungsber. Heidelb. Akad. Wiss. Math.-Natur. Kl.* **3**, 57-104.

Schwärzler, W. and Welsh, D.J.A. (1993) Knots matroids and the Ising model, *Math. Proc. Camb. Phil. Soc.* **113**, 107-139.

Schrijver, A. (1990) Tait's flyping conjecture for well connected links (preprint).

Seifert, H. (1934) Über das Geschlecht von Knoten, *Math. Ann.* **110**, 571-592.

Selman, A. (1992) A survey of one-way functions in complexity theory, *Math. Systems Theory* **25**, 203-221.

Seymour, P.D. (1981) Nowhere-zero 6-flows, *J. Comb Theory* **B 30**, 130-135.

Seymour, P.D. and Welsh, D.J.A. (1975) Combinatorial applications of an inequality from statistical mechanics, *Math. Proc. Camb. Phil. Soc.* **77**, 485-497.

Simon, J. (1977) On the difference between one and many. *Intern. Conf. Automata, Lang., Progr.*, Lecture Notes in Computer Science 52, 480-491, Springer-Verlag.

Sinclair, A. (1988) *Randomised Algorithms for Counting and Generating Combinatorial Structures*, PhD Thesis, University of Edinburgh.

Sinclair, A. (1991) Improved bounds for mixing rates of Markov chains and multicommodity flow, *Combinatorics, Probability and Computing* (to appear).

Sinclair, A.J. and Jerrum, M.R. (1989) Approximate counting, uniform generating and rapidly mixing Markov chains, *Information and Computation* **82**, 93-133.

Sokal, A.D. (1989) Monte Carlo methods in Statistical Mechanics; Foundations and New Algorithms, *Lecture Notes: Troisieme Cycle de la Physique en Suisse Romande, Semestre d'ete.*

Soteros, C.E., Sumners, D.W. and Whittington, S.G. (1992) Entanglement complexity of graphs in Z^3, *Math. Proc. Camb. Phil. Soc.* **111**, 75-91.

Stanley, R.P. (1973) Acyclic orientations of graphs, *Discrete Math.* **5**, 171-178.

Stanley, R.P. (1986) *Enumerative Combinatorics*, Wadsworth and Brooks/ Cole Monterey.

Stillwell, J. (1980) *Classical Topology and Combinatorial Group Theory*, Springer-Verlag, Berlin.

Stockmeyer, L.J. (1985) On approximation algorithms for $\#P$, *SIAM J. Computing* **14**, 849-861.

Sumners, D.W. (1987a) The role of knot theory in DNA research, *Geometry and Topology, Manifolds, Varieties and Knots* (C. McCrory, T. Schifrin, eds.), New York: Marcel Dekker, 297-318.

Sumners, D.W. (1987b) The knot theory of molecules, *J. Math. Chem.* **1**, 1-14.

Sumners, D.W. (1988) The knot enumeration problem, *Studies in Physical and Theoretical Chemistry* **57**, 67-82.

Sumners, D.W. and Whittington, S.G. (1988) Knots in self-avoiding walks, *J. Phys. A* **21**, 1689-1694.

Swendsen, R.H. and Wang, J.-S. (1987) Non universal critical dynamics in Monte Carlo simulations, *Phys. Rev. Lett.* **58**, 86-88.

Sykes, M.F. and Essam, J.W. (1964) Exact critical percolation probabilities for site and bond problems in two dimensions, *J. Math. Phys.* **5**, 1117-1127.

Tait, P.G. (1988) On Knots I, II, III, Scientific Papers Vol. I, Cambridge University Press, London, 273–347.

Tarui, J. (1991) Randomized polynomials, threshold circuits, and the polynomial hierarchy, *Proc. 8th Symp. on Theoretical Aspects of Computer Science*, Lecture Notes in Comput. Sci. 480, 238-250.

Temperley, H.N.V. (1979) Lattice models in discrete statistical mechanics, in: *Applications of Graph Theory*, (eds. R.J. Wilson and L.W. Beineke), Academic Press, London, 149-175.

Temperley, H.N.V. and Lieb, E.H. (1971) Relations between the percolation and colouring problem and other graph-theoretical problems associated with regular planar lattice: some exact results for the percolation problem, *Proc. Roy. Soc. London A* **322**, 251-280.

Thistlethwaite, M.B. (1985) Knot tabulations and related topics, *Aspects of Topology* (ed. I.M. James and E.H. Kronheimer), Cambridge Univ. Press, 1-76.

Thistlethwaite, M.B. (1987) A spanning tree expansion of the Jones polynomial, *Topology* **26**, 297-309.

Thistlethwaite, M.B. (1988) On the Kauffman polynomial of an adequate link, *Invent. Math.* **93**, 285-296.

Thomason, A.G. (1978) Hamiltonian cycles and uniquely edge colourable graphs, *Ann. Disc. Math* **3**, 259-268.

Thomassen, C. (1992) The even cycle problem for directed graphs, *J. Amer. Math. Soc.* **5**, 217-229.

Thompson, C.J. (1972) *Mathematical Statistical Mechanics*, Princeton Univ. Press.

Thurston, W.P. (1988) Finite state algorithms for the braid groups, preliminary draft.

Toda, S. (1989) On the computational power of PP and $\oplus P$, *Proc. 30th IEEE Symposium on Foundations of Computer Science*, 514-519.

Toda, S. and Ogiwara, M. (1992) Counting classes are at least as hard as the polynomial-time hierarchy, *SIAM J. Comput.* **21**, 316-328.

Toda, S. and Watanabe, O. (1992) Polynomial time 1-Turing reductions from $\#PH$ to $\#P$, *Theoret. Comp. Sc.* **100**, 205-221.

Traldi, L. (1989) A dichromatic polynomial for weighted graphs and link polynomials, *Proc. Amer. Math. Soc.* **106**, 279-286.

Traczyk, P. (1991) On the index of graphs: Index versus cycle index (preprint).

Trotter, H.F. (1969) Computations in knot theory, *Computational Problems in Abstract Algebra* (ed. J. Leech) Pergamon Press, 359-364.

Turaev, V.G. (1988) The Yang-Baxter equation and invariants of links, *Invent. Math.* **92**, 527-553.

Tutte, W.T. (1947) A ring in graph theory, *Proc. Camb. Phil. Soc.* **43**, 26-40.

Tutte, W.T. (1954) A contribution to the theory of chromatic polynomials, *Canad. J. Math.* **6**, 80-91.

Tutte, W.T. (1970) On chromatic polynomials and the golden ratio, *J. Combinatorial Theory* **9**, 289-296.

Tutte, W.T. (1976) The dichromatic polynomial, *Proc. Fifth British Combinatorial Conference, Congressus Numerantium XV*, Utilitas Math., Winnipeg, 605-635.

Tutte, W.T. (1984) *Graph Theory*, Addison Wesley, California.

van der Waerden, B.L. (1941) Die lange Reichweite der regelmässigen Atomanordnung in Mischkristallen. *Z. Physik* **118**, 473.

Valiant, L.G. (1979a) The complexity of enumeration and reliability problems, *SIAM J. Comput.* **8**, 410-421.

Valiant, L.G. (1979b) The complexity of computing the permanent, *Theor. Comp. Sci.* **8**, 189-201.

Valiant, L.G. and Vazirani, V. (1986) *NP* is as easy as detecting unique solutions, *Theoret. Comp. Sc.* **47**, 85-93.

Vertigan, D.L. (1991a) The computational complexity of Tutte invariants for planar graphs, *SIAM J. Comput.* (submitted).

Vertigan, D.L. (1991b) On the computational complexity of Tutte, Homfly and Kauffman invariants (to appear).

Vertigan, D.L. (1991c) Bicycle dimension and special points of the Tutte polynomial, *J. Comb. Th.* B (submitted).

Vertigan, D.L. and Welsh, D.J.A. (1992) The computational complexity of the Tutte plane: the bipartite case. *Combinatorics, Probability and Computer Science*, **1**, 181-187

Vogel, P. (1990) Representation of links by braids: A new algorithm, *Comment. Math. Helvetici* **65**, 104-113.

Vologodskii, A.V., Lukashin, A.V., Frank-Kamenetskii, M.D. and Anshelvi, V.V. (1974) The knot probability in statistical mechanics of polymer chains, *Sov. Phys. - JETP* **39**, 1059-1063.

Waldhausen, F. (1978) Recent results on sufficiently large 3-manifolds, *Proc. Symp. Pure Math. American Math. Soc.* **32**, 21-38,

Wang, J.C. (1982) DNA topoisomerases, *Scientific American* **247**, 94-109.

Wasserman, S.A., Dungan, J.M. and Cozzarelli, N.R. (1985) Discovery of a predicted DNA knot substantiates a model for site-specific recombination, *Science* **229**, 171-174.

Welsh, D.J.A. (1976) *Matroid Theory*, London Math. Soc. Monograph **8**, Academic Press, London.

Welsh, D.J.A. (1988) *Codes and Cryptography*, Oxford Univ. Press.

Welsh, D.J.A. (1990) The computational complexity of some classical problems from statistical physics, *Disorder in Physical Systems*, Oxford Univ. Press, 307-323.

Welsh, D.J.A. (1991) On the number of knots and links, *Colloq. Math. Soc. Janos Bolyai*, **60**, 713-718.

Welsh, D.J.A. (1992a) The complexity of knots, *Ann. Disc. Math* **55**, to appear.

Welsh, D.J.A. (1992b) Knots and braids; some algorithmic questions, *Contemp. Math., Amer. Math. Soc.*, to appear.

Welsh, D.J.A. (1992c) Percolation in the random cluster process and Potts model, *J. Phys A.* to appear.

Whitehead, J.H.C. (1937) On doubled knots, *J. Lond. Math. Soc.* **12**, 63-71.

Whitney, H. (1932) A logical expansion in mathematics. *Bull. Amer. Math. Soc.* **38**, 572-579.

Whitten, W. (1987) Knot complements and graphs, *Topology* **26**, 41-44.

Whittington, S.G. (1992) Topology of polymers, *AMS Proc. Symp. Appl. Math.* **45**, 73-95.

Wierman, J.C. (1981) Bond percolation on honeycomb and triangular lattices, *Advances in Applied Probability* **13**, 293-313.

Witten, E. (1989) Gauge theories and integrable lattice models, *Nucl. Phys. B* **322**, 629-697.

Woodall, D.R. (1992) A zero-free interval for chromatic polynomials, *Discrete Math.* **101**, 333-341.

Wu, F.Y. (1982) The Potts model, *Rev. Mod. Phys.* **54**, 235-268.

Wu, F.Y. (1984) Potts model of magnetism, *J. Appl. Phys.* **55**, 2421-2424.

Wu, F.Y. (1988) Potts model and graph theory, *J. Stat. Phys.* **52**, 99-112.

Wu, F.Y. (1992) Knot theory and statistical mechanics, *Rev. Mod. Phys.* **64**, 1099-1131.

Yajima, T. and Kinoshita, S. (1957) On the graphs of knots, *Osaka Math J.* **9**, 155-163.

Yamada, S. (1987) The minimal number of Seifert circles equals the braid index of a link, *Inv. Math.* **89**, 347-356.

Zaslavsky, T. (1975) Facing up to arrangements: face count formulas for partitions of spaces by hyperplanes. *Memoirs Amer. Math. Soc.* **154** (American Mathematical Society).

Zaslavsky, T. (1992) Strong Tutte functions of matroids and graphs, *Trans. Amer. Math. Soc.*, to appear.

Index

Printed in the United States
By Bookmasters